PREDICTING THE UNPREDICTABLE

PREDICTING THE UNPREDICTABLE

The Tumultuous Science of
Earthquake Prediction

Susan Hough

PRINCETON UNIVERSITY PRESS   *Princeton & Oxford*

Copyright © 2010 by Princeton University Press

Published by Princeton University Press, 41 William Street, Princeton, New
Jersey 08540

In the United Kingdom: Princeton University Press, 6 Oxford Street,
Woodstock, Oxfordshire OX20 1TW

Library of Congress Cataloging-in-Publication Data

Hough, Susan Elizabeth, 1961–
Predicting the unpredictable : the tumultuous science of earthquake
  prediction / Susan Hough.
p.   cm.
Includes bibliographical references and index.
ISBN 978-0-691-13816-9 (hardcover : alk. paper)   1. Earthquake prediction.
  I. Title.
QE538.8.H68 2010
551.22—dc22   2009008380

British Library Cataloging-in-Publication Data is available

This book has been composed in Bembo and Berthold Akzidenz Grotesk

Printed on acid-free paper. ∞

press.princeton.edu

Printed in the United States of America

10 9 8 7 6 5 4 3 2 1

Contents

PREDICTING THE UNPREDICTABLE

CHAPTER 1

Ready to Rumble

> What is most tragic is that the collective genius
> of all of these experts, combined with the sensors and
> satellite observations and seismographic data and all the
> other tools of science and technology, could not send the
> important message at the key moment: Run.
> Run for your lives.
>
> —JOEL ACHENBACH, *Washington Post*, January 30, 2005

At the beginning of 2005, U.S. Geological Survey geophysicist Bob Dollar was keeping a routine eye on data from the local Global Positioning System (GPS) network in southern California, and something caught his eye. A small army of GPS instruments throughout California tracks the motion of the earth's tectonic plates; the movement of the North American Plate south relative to the Pacific Plate as well as more complicated, smaller-scale shifts. Plates move about as fast as fingernails grow; like fingernails, the movement is not only slow but also steady (fig. 1.1). But it seemed to Dollar that a group of stations out in the Mojave Desert and some in the San Gabriel Valley northeast of central Los Angeles had started to take a bit of a detour from their usual, steady trajectories.

When one uses GPS data to determine precise locations the results always reveal some flutter, the consequence of measurement imprecision or data processing complications. Knowing this, Dollar did not jump out of his chair. But, interest peaked, he continued to keep his eye on the results, waiting for the apparent detours to prove to be part of the usual noise.

They didn't. After a couple of months of watching and waiting, the

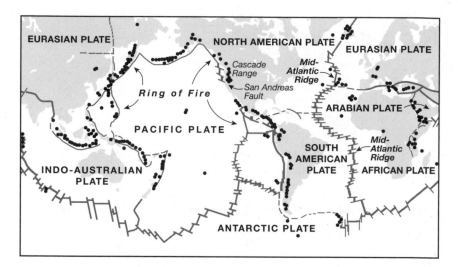

Figure 1.1. The earth's major tectonic plates. The so-called Ring of Fire includes both transform faults such as the San Andreas Fault in California and several major subduction zones around the Pacific Rim. The dots in the figure indicate active volcanoes. (Image courtesy of USGS.)

hiccups took shape, defining what Dollar calls "hockey-stick curves." Which is to say, the data from a number of stations, formerly tracking along straight lines, had bent abruptly and were now tracking along different lines. It was enough to get any self-respecting earthquake scientist's attention. Dollar started to think that he "might be looking at something important." What the results meant, he wasn't sure. At a minimum, departures from steady GPS trajectories are unusual, and therefore interesting. But several lines of evidence suggest that this kind of anomaly, essentially abrupt and unusual warping of the earth's crust, could be a harbinger of a future large earthquake.

Earthquake predictions emerge from the pseudo-science community not unlike locust plagues in the desert: not like clockwork, exactly, but often enough. In these cases seismologists can speak to the media with confidence. At best, these sorts of predictions rely on methods that might (emphasis here on *might*) have an underlying shred of validity—for example, the notion that tidal forces might influence earthquakes—but have never proven useful for reliable earthquake prediction. At worst they are total hooey. But every once in a while the earth puts out signals that get scientists' attentions, leading us to wonder, is the Big One coming?

Arguably the biggest unanswered question in earthquake science is this: what, if anything, happens in the earth that sets a big earthquake in motion? The answer might be, "nothing." Earthquakes might pop off in the crust like popcorn kernels, at a more-or-less steady rate, leaving us with no way to tell which of the many small earthquakes will grow into the occasional big earthquake. If this is the case, too bad for earthquake prediction. But at least some theories and bits of evidence suggest that earthquakes might have a detectable launch sequence.

The last great earthquake in California was over one hundred years ago. The 1906 San Francisco earthquake was recorded on a handful of early seismometers around the world, and geodetic surveying measurements made before and after the quake led directly to one of the most fundamental tenets of earthquake science. The theory of elastic rebound describes how earthquakes happen as a consequence of stress accumulation. The theory of plate tectonics, developed a half-century later, explains how and why stress accumulates. In short: plates move, the edges stay locked, the surrounding crust warps, eventually the edges (faults) move abruptly to catch up. But if the earth sent out any subtle signals that the 1906 earthquake was on the way, they were lost forever, no instruments in place to capture them.

In recent years, scientists have developed and deployed increasingly sophisticated instruments to capture signals from the earth, not only earthquake waves but also minute warping of the crust. If any subtle signals are generated prior to large earthquakes, these instruments stand ready and waiting. Data have been recorded prior to a number of recent moderate, magnitude 6–7, earthquakes in California. And they have revealed no sign of precursory signals. This negative result has led some earth scientists to conclude that there is nothing to be revealed; that, in effect, earthquakes have no launch sequence. But we have yet to see the likes of the 1906 San Francisco earthquake caught red-handed by a dense, close-in array of modern instruments. So seismologists are left to wonder what, if anything, the instruments will reveal when the next Big One strikes. Thus when instruments reveal something outside the ordinary, we are left to wonder, could this be *it*?

In early spring of 2005 Dollar brought his hockey-stick curves to the attention of local GPS gurus. They were not immediately impressed. One geophysicist confessed that her first thought upon seeing the curves was, "What did we do wrong?" Scientists who study GPS and related data are not simply inclined toward self-doubt; they have learned to not get too excited about apparently unusual signals. GPS instruments essentially record time signals from satellites, and scientists use these signals to determine locations. The processing is notoriously complicated and capricious for a number of reasons, including the fact that the raw data have to be corrected very carefully to account for the orbits of the satellites. The results that Dollar had been looking at were from rapid—essentially quick and dirty—solutions. When researchers analyze GPS data for scientific investigations the raw data are processed more carefully. Not uncommonly, glitches in quick solutions disappear when more sophisticated processing is done.

Dollar's hockey sticks refused to flatten out. Eventually the results came to the attention of other colleagues, not GPS gurus but rather seismologists, and they took note. Where Dollar had been thinking he might be looking at something important, some seismologists wondered if they were looking at something alarming. By this time the signals had lasted long enough that local GPS experts were also convinced they were more than a glitch. Several top seismologists sprang into action. Meetings were held. Memos were written. Blood pressures rose.

Earthquake science is not a good business to be in if one is a control freak at heart. As a seismologist, one's career hangs at the mercy of infrequent and unpredictable events. We pursue research plans knowing that, at any moment, those plans could be blown out of the water by an earthquake that will consume all of our time and energy for months, if not years. Most of the time such thoughts can be pushed to the back of one's mind. Every so often, it's not so easy. Spring of 2005 was one of those times for earthquake scientists in southern California. Along with everyone else we had horrific images of the December 26, 2004 Sumatra earthquake and tsunami freshly seared onto our minds. Nor did it help that a growing body of evidence seemed to indicate that it has been a very long time, maybe too long, since the last Big One in southern California. Of particular concern, the San Andreas and San Jacinto faults in the southernmost one-third of the state, roughly from Palm Springs to near the Mexico border, have remained stubbornly locked for over three hundred years. Farther north, between San Bernardino and central California, the San Andreas Fault last broke way back in 1857 (fig. 1.2). This does not tend to be a source of comfort. The best geological evidence suggests that big quakes occur on both of these fault segments about every 150 to 300 years, maybe less. We also can't rule out the possibility that both segments of the southern San Andreas could unzip in a single earthquake, what we sometimes call a wall-to-wall rupture. If 1857 was Big, a wall-to-wall rupture of the southern San Andreas Fault would be Bad.

A REALLY BAD ONE?

Scientists categorically discount the possibility that the San Andreas Fault could rupture stem to stern: a single massive break running nearly the entire length of California. We do not believe a rupture to continue to propagate through a middle section of the fault that does not lock up, but along which movement occurs via steady creep. Continuing an earthquake through this section would be a bit like trying to propagate a crisp tear through a soggy patch of newsprint. As assumptions go this one is well founded. But well-founded assumptions have been wrong before.

Results from recent investigations of the southern San Andreas Fault have found their way into scientific journals and, from there, into mainstream publications. Newspapers sometimes add their own exclamation points. In late 2006, one particularly memorable headline

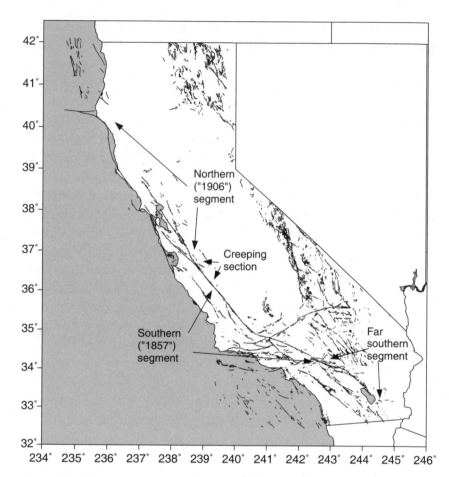

Figure 1.2. The San Andreas Fault in California. Other faults in the state are also shown.

splashed far and wide, "Southern San Andreas Fault Ready to Explode!"

Concern for the southern San Andreas is scarcely new. The *Nature* paper that sparked the 2006 headlines used a new technology (synthetic aperture radar) to confirm and explore in detail a result that had been known for years, if not decades. When the curious GPS signal cropped up in the spring of 2005, every earthquake scientist in southern California knew that it had been a long time since the last Big One. But what to make of the signal? Had the complex data processing somehow gone wrong? If the earth itself had hiccupped, what did it mean?

And at what point would it be responsible to communicate concern to the public?

Earthquake scientists have come by caution the hard way. The two most notorious earthquake prediction scares in California during the twentieth century were based on apparent signs of ominous warping that turned out to be the consequence of imprecise data and/or faulty interpretation. In both of the earlier cases, the apparently ominous warping had been revealed with traditional surveying techniques, whereas the 2005 signal had been measured with modern GPS instruments. But the parallels alone were enough to give the judicious earth scientist pause. At the same time, the doubt nags at the back of one's mind: if, as seismologists, we are seeing signals that leave us concerned, is it responsible to *not* communicate that concern to the public? And the doubt that nags more seriously: what if the Big One strikes while we continue to grapple with the question of going public?

Most seismologists are not quite so clueless as to admit in public that we like earthquakes. Even if it is partially true it sounds wrong. We might be geeks but we are not ghouls. When journalist Joel Achenbach commented on the fundamental communication failure that contributed to the staggering death toll of the December 26, 2004 Sumatra earthquake and tsunami, some earth scientists took exception to the perceived intimation that scientists don't try to translate knowledge into effective communication and hazard mitigation. For most of us who work on hazard-related science Achenbach's words weren't accusatory but rather poignant. We do try. It isn't easy. It especially isn't easy when one struggles to communicate the appropriate message based on incomplete and ambiguous information. To sound alarm sirens when a tsunami wave is approaching, this is a logistical challenge. To sound an alarm when we see an unusual signal that we don't fully understand, this is a challenge that cannot be solved with monitoring equipment and T-1 lines and sirens.

Investigations of GPS data ordinarily proceed at an unhurried pace. It takes years if not decades to collect the data in the first place. And like any scientific investigation it typically takes months to analyze the data, write up the results, and many more months for a paper to navigate the peer-review process. In the spring of 2005, a small group of scientists at the U.S. Geological Survey and the Jet Propulsion Laboratory didn't have time. They had heartburn.

The first order of business was to check and recheck the basic processing of the GPS data. A handheld GPS receiver can track your position accurately enough to navigate on city streets, but geophysical investigations, which require millimeter-scale accuracy, are a different ball game. In addition to satellite orbit corrections, when one tracks the position of a GPS instrument, one has to ask the question, motion relative to *what*? Rephrasing the question in scientific terms, what is the reference frame? This might sound like a simple question; it isn't. Using the best-available methods to process data, Tom Herring at MIT showed that part of the apparently unusual signal resulted from a subtle reference-frame issue. The magnitude 9.3 Sumatra earthquake was so enormous that it had caused small readjustments all over the planet. Taking those readjustments into account, the apparent anomaly in the Mojave Desert disappeared. The so-called San Gabriel anomaly, however, did not go away. In fact, it was revealed a fairly simple, broad and significant uplift of the crust.

Convinced that the San Gabriel signal was real, GPS experts turned to the next question: what had caused it? Was it a sign that the crust was warping suddenly (read: ominously) around the buried Whittier Fault just east of central Los Angeles? Or could the signal be hydrological in origin, the consequence of changes in ground water?

January of 2005 was a memorable month for southern Californians. Between December 27, 2004 and January 10, 2005, downtown Los Angeles received a whisker shy of seventeen inches of rain, three inches more than the city receives in an average year. Some foothill communities got soaked far worse. The rains were not only epic, they were historic. There is an unwritten law in southern California, understood by the public and agreed to by the gods: it does not rain on the Rose Parade. In 2005, for the first time in a half-century, the gods failed to hold up their end of the deal.

By 2005 scientists understood that groundwater can cause the ground to move up and down, both via natural recharge of aquifers during the rainy season and as a consequence of groundwater extraction during dry months. Usually the recharge process is gradual. But usually Los Angeles doesn't get seventeen inches of rain in fourteen days.

Looking at the San Gabriel anomaly scientists fell in one of two camps: those who were pretty sure it had been caused by rainfall, and those who weren't. It was really only a matter of educated opinion, how scientists sorted themselves into these camps, although the GPS experts generally remained more sanguine than—and occasionally irked by—some of their seismological colleagues. But whatever their hunches might be, GPS experts knew they had to work, and work fast, to come up with a definitive answer. Or, if not a definitive answer, at least one that settled the issue beyond reasonable doubt.

A team of scientists at the U.S. Geological Survey and Jet Propulsion Lab first worked to explore the extent of the warping using the most careful, sophisticated data processing. They confirmed that a broad swath of ground had moved upward by as much as four centimeters—not quite two inches. They then asked, could this warping be explained by a build-up of strain on a buried fault? The answer was, not easily. If strain were to suddenly build up on a fault, one would expect the warping to be centered on that fault. On the one hand the extent of the San Gabriel Valley anomaly did not coincide with any one fault. On the other hand, independent estimates of groundwater elevation—the depth of water within the earth's crust—revealed an abrupt increase that coincided with the timing of the anomalous GPS signal. Further, by late spring of 2005, both the groundwater and the GPS trends had started to reverse; in effect, the San Gabriel Valley began to exhale, about as close as one could get to a smoking gun pointing to groundwater as the cause of the inhale.

By late summer of 2005 the scientific community was able to exhale as well. The sense of urgency defused, science resumed its usual course. Talks were presented at scientific meetings in late 2005 and early 2006. The definitive paper appeared in the prestigious *Journal of Geophysical Research* in early 2007. A press release went out when the paper was published, anticipating some public interest in the discovery that the San Gabriel Valley had been swelled upward temporarily by rainfall. It was a scientifically interesting result, also an impressive demonstration of the sophistication of modern instruments. A couple of local newspapers ran brief stories; otherwise, news media ignored what they sized up as news of little consequence.

The press release did not say that this was an earthquake prediction scare that never happened. Even if it had, it is unlikely that the media would have paid much attention. The dog that doesn't bite is not news. For the credibility of earthquake science in the public arena, it is unfortunate that, while failed predictions are big news, nobody ever hears about judicious decisions not to step forward with premature concern. Had concern about the anomaly leaked—or been communicated—to the media in early 2005, it would have been big news.

In fact, as a later chapter will discuss in more detail, a different earthquake prediction story had hit the fan in the spring of 2004. A team of researchers at the University of California at Los Angeles went public with a prediction that a magnitude 6.4 or larger quake would strike the southern California desert by September 5, 2004. This prediction, based on apparent past patterns of small and moderate earthquakes preceding previous large earthquakes in California and elsewhere, failed. Not only did no large earthquake strike the target region during the prediction window; if anything the region remained unusually quiet throughout 2004. If a person didn't know better, he or she could start to think that the planet is determined to instill humility in scientists who dare to believe they have unlocked her secrets.

But we do think we know a thing or two about earthquakes. We know that in a place like California, it isn't a matter of if the next Big One will strike, but only when. We know that big earthquakes on the San Andreas don't strike like clockwork, but neither are they completely random. We know that it has been rather a long time since 1857, and even longer since the last Big One on the southernmost San

Andreas Fault. The predictions and headlines and worrisome signals come and go, but a groundswell of concern remains. And with the concern, the questions. Is the San Andreas Fault, along with other key faults that have been quiet for a long time, ready to rumble? With a history of predictions that inspires caution in any judicious earthquake scientist, how do we weigh caution against concern? And if the community of earthquake science professionals struggles with these issues, what should the public make of the whole mess?

For nearly a century scientists as well as residents of southern California have lived under a sword. We know that a very big earthquake will strike the region some day; we don't know if that day is tomorrow or fifty years from now. It is therefore no surprise that the history of earthquake prediction research, in the United States in particular, is inexorably intertwined with the history of earthquake science in southern California.

But what's up with earthquake prediction research, anyway? Have scientists made any progress since the 1970s, when many experts went public with their belief that reliable earthquake prediction was just around the corner? What of the persistent belief held by many outside of science, that animals can sense impending earthquakes? or that earthquakes are triggered by lunar tides? Didn't the Chinese successfully predict a big earthquake back in the 1970s? If they predicted an earthquake thirty years ago, why was there no warning in advance of the deadly Sichuan earthquake of 2008?

The story of earthquake prediction is a story about science, but not only that. It is a story about what happens when the world of science collides with an outside world that has a life-and-death stake in research that continues to be a work in progress. It is a story that pulls back the curtain to reveal the inner workings of science; a business that is often far more messy, and far less divorced from politics as well as personality, than the public realizes and scientists like to believe. It is a story that does not end—that might not ever—the way we want it to end. It is a story we can't put down.

CHAPTER 2

Ready to Explode

"General Earthquake or Series Expected"
—Newspaper headline, *Sheboygan Press*,
November 16, 1925

*I*f the San Gabriel anomaly was the dog that didn't bite, it was descended from a toothier breed. Since the early twentieth century southern California has been a hotbed for not only earthquake science but also earthquake prediction; not only for earthquake prediction research but also for earthquake prediction fiascos.

The beginnings of earthquake exploration in southern California date back to 1921, when geologist Harry Wood convinced the Carnegie Institute to underwrite a seismological laboratory in Pasadena. Looking to record local earthquakes, Wood teamed up with astronomer John August Anderson to design a seismometer that could record small local rumblings. By the late 1920s a half-dozen "Wood-Anderson" seismometers were in operation throughout southern California. In 1928 the lab hired a young assistant with a physics degree to start to analyze seismograms: Charles F. Richter. Five years later his formulation of a first-ever magnitude scale provided the basis for the first-ever modern earthquake catalog—arguably the start of modern network seismology.

Before network seismology even began, geologists' attentions had turned to southern California; among them, Bailey Willis. Born in New York two months after the great 1857 Fort Tejon, California earthquake, Willis made his way from engineering to geology, landing at Stanford University as a professor and chairman of the department in 1915. Although no longer a young man, Willis had tremendous physical as well as intellectual energy. His scientific career—including field investigations that took him to the far corners of the globe—could have easily filled one lifetime. But the son of poet and journalist Nathaniel Parker Willis was not destined for a one-dimensional life. He had five children, three with second wife Cornelia following the death of his first wife. He was an enthusiastic and gifted public speaker. His stirring lectures occasionally received standing ovations, the likes of which do not happen every day in scientific circles. He pursued watercolor painting as a serious avocation. He excelled at cabinetry (fig. 2.1).

Willis had not been active in seismology at the time of the great 1906 San Francisco earthquake. Nor did he join efforts in the immediate aftermath of the earthquake to launch the Seismological Society of America (SSA), an organization whose primary mission was and remains promotion of improved understanding of earthquakes and earthquake hazard. Having gotten off to a somewhat slow and jerky (so to speak) start during its first decade or so, under the leadership of Stanford president John Branner, the society had taken root and gained a measure of momentum by the time Willis arrived at Stanford. Landing in the Bay Area in the aftermath of 1906, Willis's boundless intellectual curiosity and energy drew him naturally to earthquake studies, and the SSA. By 1921 Willis assumed the reins—and firm control—of the fledgling society.

Conceived as a scientific organization dedicated to safety as well as science, the society was probably destined from the beginning to veer in a more purely scientific direction. There is no question that the organization eventually evolved in this way; with earthquake engineering falling to specialized engineering societies. Even the earliest issues of the *Bulletin of the Seismological Society of America* include far more science than engineering. But when elected by his peers to be president of the society in 1921, Willis brought not only a geologist's energy and passion but also an engineer's sensibilities to the job. Even at that early date Willis and others knew that proper engineering could go a long way toward guaranteeing safety during the future large earthquakes that, as geologists knew by then, were inevitable.

Figure 2.1. Bailey Willis. (Photograph courtesy of Clay Hamilton.)

It did not escape Willis's attentions that early issues of the *Bulletin* had a strong scientific bent. As president of the society he formulated a plan to publish one special issue of the *Bulletin* devoted entirely to earthquake engineering. Willis wanted this issue to include a map showing all known faults in the state. He knew that such a map would reveal a state crisscrossed by active faults; he knew it would show that virtually no corner of the state is free from earthquake hazard. Nowadays, a geologist armed with modern computer tools can produce a map like this in the time it takes to read this chapter. In 1921 it was no small feat. For starters, one first had to figure out where the faults were.

Some of California's faults are more obvious than others. Geologists had recognized only bits and pieces of the San Andreas Fault prior to 1906, but the San Francisco earthquake left a jagged scar along the fault from San Juan Bautista, about ninety miles south of San Francisco, to Point Arena. Geologists set out to trace this fresh 185-mile scar, among them Harold Fairbanks. Although less well known today in earthquake circles than colleagues whose names became more indelibly attached to the 1906 earthquake, Fairbanks followed the scar to its southern terminus. Then he kept going. Working in a part of California so remote that, still today, it remains the playground of antelope and mountain lions, Fairbanks continued to follow the fault, eventually mapping nearly its full extent (fig. 2.2).

Fairbanks and his team lost the trail in the rugged San Gorgonio Pass region east of San Bernardino. He can be forgiven for this. Even with the ability to view faults from the air, which provides a much clearer view of large-scale geological features, geologists still struggle to identify the precise location of the fault(s) through this pass. Some question whether the San Andreas continues uninterrupted through this geological knot. Fairbanks observed, more or less correctly, that, "it is probable that a fault continues on still farther along the mountains lying north of the Salton Basin." In fact the San Andreas extends and ultimately dies out along the eastern edge of the Salton Sea. But it wasn't a bad guess.

Meanwhile, other geologists had been busy mapping other faults. In particular, the 1892 discovery of oil in the Los Angeles region sent industry geologists scurrying to understand the geological structure of the Los Angeles basin. By 1921 they had mapped many of the important faults in the region. Understanding subsurface geology remains the bread and butter of the petroleum industry. Today oil companies use sophisticated techniques to develop detailed images of the subsurface, not unlike a CAT scan of a human body. Oil companies are un-

Figure 2.2. The San Andreas Fault through the Carrizo Plain, seen from the air. (USGS photograph by Robert Wallace.)

derstandably not always eager to share proprietary data. In 1921, local industries were willing to share their surface fault maps with Harry Wood. Compiling mapped faults was itself no small task in an era when mapping had to be done by hand, without the benefit of aerial photos. Wood further took time to conduct his own fault-finding expeditions in parts of the Los Angeles region not covered by industry maps.

Willis ultimately fell short in his hope for a special issue of the *Bulletin*, but, fueled by Wood's tenacity as well as his own, the state fault map began to take shape. Willis compiled information on faults in the Bay Area. To pull together information for central California Willis contracted with a Stanford student, his son Robin, who went on to a career in the oil industry. By the fall of 1922 Willis had his map. Later that year he unveiled it at a meeting of the American Association for the Advancement of Science. The task remained, however, to print and distribute the map. The *Bulletin of the Seismological of America* provided an obvious and willing vehicle, but the society lacked the resources required to pay for printing and distribution.

Not a man easily deterred, Willis turned himself into a one-man fund-raising campaign and membership drive. He recruited leading businessmen as well as scientists to join the society, in just a few years nearly doubling the 1920 membership of 307. He further appealed to San Francisco businessmen, raising almost $1,200 for the society. This fell short of the $10,000 he had sought to raise to publish the map and establish the society on a firm footing. Not for the first time the private Carnegie Institution was enjoined to step in. Having underwritten publication costs for an exhaustive report on the 1906 San Francisco earthquake in addition to financing the seismological laboratory in Pasadena, they contributed another $5,000 to the society in response to Willis's appeal. The fault map was published in the *Bulletin* in late 1923. It was further advertised in a number of other journals, and distributed via libraries and civic organizations. Willis's vision had borne fruit. Not only scientists but also businessmen and the public could now see with their own eyes that California is earthquake country. It was an impressive achievement. Geologists have learned a lot about faults since the early decades of the twentieth century, but to the

modern trained eye, Willis's map bears a remarkable resemblance to
modern fault maps (fig. 2.3).

Thus by early 1925 the first-ever California earthquake awareness
campaign was well underway. And seismologists had just begun to record local earthquakes, with the hope that their locations and patterns
would provide clues to the locations of future large earthquakes.

The 1906 earthquake had commanded the attentions of the public
as well as the scientific community. Thanks to the efforts of early seismologists including Harry Wood, California was home to a thriving,
dynamic community of earthquake scientists. These scientists had made
impressive strides in a few short years, gaining a better understanding
of both earthquakes and faults. Yet Willis's dream of a publication
aimed specifically at improving building safety had fallen by the wayside. His plans for a Build for Security crusade had sputtered and died.
In the end he solicited a single article, "Earthquake-Proof Construction," from Japanese architect, engineer, and professor Tachu Naito, regarded as a founding father of earthquake-resistant design.

A less tenacious individual might have been content to revel in his
success. By early 1925, Willis was more convinced than ever that scientific information alone was not enough. To make a difference, Willis
realized, scientists had to become advocates. They had to become
spokesmen.

Scientists today have an appreciation of the importance of outreach
and education. Not all research scientists are happy to take time away
from their work to spend time talking with the media and public, but
more are willing to do so today than in Willis's day. Reading early
twentieth-century California newspaper articles about earthquakes,
one is struck by the small number of scientists whose names appear.
Harry Wood. Charles Richter. Bailey Willis. Richter, who began talking with the media in the late 1920s, would become the leading earthquake spokesman through much of the twentieth century. Before
Richter, there was Willis. Willis not only answered journalists' questions, he also talked to anyone who would listen about the need for
effective earthquake provisions in building codes. He took his speaking talents on the road.

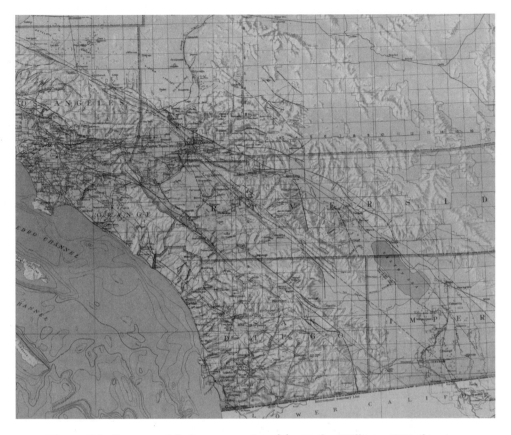

Figure 2.3. Excerpt of fault map produced by Bailey Willis in 1923 showing faults in the greater Los Angeles region.

One might wonder: what motivates a man like Willis to embark on such a crusade? Were his motivations entirely noble, or did he perhaps enjoy the limelight? Certainly his successor, Charles Richter, would find himself dogged by accusations of this. Following Willis's death at the age of ninety-one, his closest colleagues at Stanford wrote, "The personality of this extraordinary man was one of his engaging assets. People were delighted by his words and manner, his humor and lack of ostentation. He made a strong impression upon his students, and he left them with a feeling of warmth and friendliness."

In 1947 Willis penned a memoir, *A Yanqui in Patagonia*, focused mainly on his leadership of field campaigns in Argentina and Patagonia between 1910 and 1914. It is a tale of adventure and accomplishment, with little hint of overabundant ego. If darker motivations lurked beneath the surface, he hid them well.

In the 1920s, Willis's not inconsiderable powers of persuasion ran headlong into the state's not inconsiderable business interests. In the San Francisco area nobody could deny the reality of earthquake hazard, but businesses could and did try to deny the necessity of stringent building codes that would raise significantly the cost of construction. They could and did argue against the need for prohibitively expensive insurance rates. After all, most of the damage in 1906 had been caused by fire, not shaking. San Francisco also put itself back together in record time, within a few short years presenting itself to the world as back in business, better than before.

To the south, by the mid-1920s Los Angeles had started to grow. Drawn to the area by opportunities such as oil and the entertainment industry, not to mention a congenial climate, over 500,000 people—almost all of them transplants—lived in the city by 1920. Within a decade that number had more than doubled. Business was good, and business leaders rushed to paint the region in nothing short of idyllic terms: not only free from serious earthquake risk but also immune from the violent storms that ravage other parts of the country.

By 1925 business leaders had Willis's fault map to contend with, but that was easily brushed aside. The San Andreas Fault did not run through Los Angeles; it ran along the northern flank of the San Gabriels, a mountain range away from the city. Those faults mapped by oil company geologists within the Los Angeles Basin? They were dismissed as inactive.

For the purposes of earthquake awareness, stirring rhetoric and compelling scientific arguments only get you so far. To really capture public attention, nothing brings home the danger of future earthquakes more effectively than a big earthquake in the recent past. Big and/or damaging moderate earthquakes leave not only the public but also scientists feeling antsy, wondering what might happen next. To

Figure 2.4. Damage from the 1925 Santa Barbara, California earthquake. (USGS photograph.)

some extent this reflects a rational consideration of scientific results: we know that earthquakes tend to beget other earthquakes. To some extent it reflects adrenaline.

Willis's crusade thus received a modest shot in the arm on the morning of June 29, 1925, when a powerful earthquake struck near Santa Barbara. The earthquake might have benefited Willis's cause but it didn't do Santa Barbara any good; thirteen people were killed. Although really only a moderate earthquake at magnitude 6.3, the event reminded Californians that it doesn't take a great earthquake to cause damage and loss of life. The shock highlighted the vulnerability of substandard construction, taking a particularly heavy toll on old buildings in the downtown business district (fig. 2.4). No firestorm had erupted this time; the damage had clearly been a consequence of the shaking. Finally, the earthquake provided an exclamation point to Willis's arguments that northern California does not have a monopoly on earthquake hazard.

But, while dramatic and deadly, the Santa Barbara earthquake was too removed from the Los Angeles region to leave any real—or at

least, any lasting—impression on local leaders, let alone on state officials. The impression on Bailey Willis was more pronounced. *This was
what he had been talking about.* A damaging earthquake on one of California's many faults, causing damage and loss of life that was almost
entirely preventable. Willis sprang into action. In the aftermath of the
quake he persuaded the city councils in Santa Barbara and his home-
town of Palo Alto to require earthquake-resistant construction in local
building codes. If he hoped to get broader, statewide attention, he
again fell short.

It isn't hard to imagine the measure of frustration that Willis must
have felt in the aftermath of the Santa Barbara earthquake. On top of
everything else, he had recently become aware of scientific results that
fueled his concern for future earthquakes in southern California in
particular. In the early 1920s the Coast and Geodetic Survey had run
surveying lines across the southern San Andreas Fault. A comparison
with an earlier survey revealed that strain was building along the fault at
the astonishing rate of twenty-four feet in thirty years (fig. 2.5). That is,
the earth's crust on both sides of the fault was being pushed in opposite
directions, but the fault itself remained locked and so the surrounding
crust was being increasingly warped. Rocks have only limited ability to
warp; eventually, Willis knew, the fault had to move. And if twenty-four
feet of movement had been stored up in just thirty years, it seemed in-
escapable that the fault would move sooner rather than later.

How much sooner, Willis knew that scientists couldn't say. But his
concern ratcheted upward and along with it, his rhetoric. In November of 1925 he told the *Daily Palo Alto*, "No one knows whether it will
be one year or ten before a severe earthquake comes, but when it does
come it will come suddenly, and those who are not prepared will suffer." The story garnered national media attention. Readers in Sheboygan, Wisconsin read that, "Any seismologist would be surprised if 10
years go by without an earthquake in southern California greater than
the recent Santa Barbara quake." Whether they misconstrued Willis's
words or read them correctly remains unclear, but some journalists
quickly recast his statement in more definitive terms. By the time the
story reached *Time* magazine, Willis's statements had evolved into a
full-fledged prediction. The magazine informed readers that Willis had

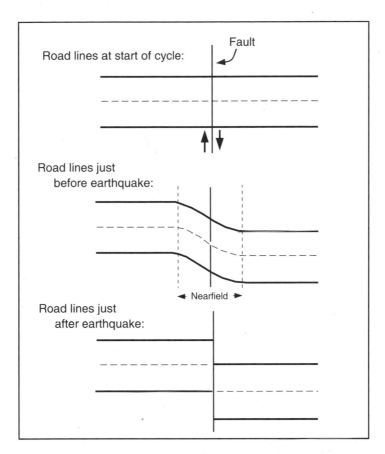

Figure 2.5. Illustration of elastic rebound theory. Road lines crossing a fault will stay pinned right at the fault. Well away from the fault the crust continues to move. The curved road lines indicate a zone of elasticity in which the crust becomes strained as the fault remains locked. When an earthquake occurs the fault and the broader strained region lurch forward to catch up. Surveying data from the early 1920s purportedly showed that twenty-four feet of warping had taken place across the San Andreas Fault in just thirty years.

"declared that within the next ten years Los Angeles will be wrenched by a tremor worse than that of San Francisco."

Over the next few years Willis walked a fine line. In his own statements to the media and business leaders he continued to say, "probably." Yet he apparently did not take pains to correct those who cast his statements as a more definitive prediction. In 1927 he brought his stump speech before the Board of Fire Underwriters of the Pacific and the National Board of Underwriters in New York. Insurance rates in Los Angeles immediately skyrocketed. These events apparently served as the impetus for Robert T. Hill, widely believed to have had insurance industry backing, to write a benignly titled book, *Southern California Geology and Los Angeles Earthquakes.* Chapter one of the book, "Southern California Attacked," focused specifically on Willis's "predictions." Any question about where Hill was coming from are dispelled by page two: "The author of these fearsome predictions is a scientific man of standing but in my opinion he has . . . done more to discredit science in the minds of the people of this vicinity than any other incident in recent years." Hill further stated outright, "It is my object in this report to refute his inconsiderate, inaccurate, and damaging prophecies."

The fact that "damaging prophecies" had been made for the city of Los Angeles by a San Francisco man only added insult to injury.

In subsequent chapters Hill proceeded to present the first detailed description of faults in the Los Angeles region. Then he went on to summarize arguments that they were all harmless relics, "things of past geologic time." By the end of the book he stepped back onto his soapbox. "Southern California is a region in which the fractures are broadly distributed," he concluded, "so that intense accumulations and releases of strain are not so strongly probable as where the fractures are closely spaced." Of the conspicuous fault running across the west side, Hill wrote, "The authorities of the University of California at Los Angeles, by permitting the new buildings at Westwood to be built almost astride of this line have shown their lack of fear of harm from it." Hill seems to have missed the sometimes-fine distinction between "lack of fear" and "good sense."

The book was in fact a seminal contribution to the understanding of faults in southern California, a first-ever comprehensive survey. "How he could have been so spectacularly right and so spectacularly wrong on the same pages is remarkable," observed Caltech geologist Clarence Allen.

By the time Hill set out to write his remarkable little book, the alarming results that had fueled Willis's concerns had been refuted—by none other than the chief of the organization responsible for collecting the data. Captain William Bowie, head of the Coast and Geodetic Survey, had been aware as early as 1924 that the alarming surveying results could not be double-checked until a subsequent survey was completed. In 1924 he expressed confidence in the soundness of the results. By the fall of 1927, however, the resurvey had been completed and the data analyzed, and twenty-four feet had become five feet. Although imprecise by modern standards, the new result came much closer to modern estimates of how fast strain, or energy, builds up along the southern San Andreas Fault.

Hill promptly made a refutation of the prediction to a meeting of the Building Owners and Managers Association; a transcript of the speech was distributed by Hill's backers, and made its way to local newspapers. Hill also spoke out to journalists, casting aspersions on not only Willis's science but also his motivations: "It is generally believed," he said, "that Dr. Willis' service to the fire insurance underwriters was substantially rewarded."

By this time a backlash against Willis's outspoken ways had arisen within the Seismological Society to which he had early brought such energy and passion. Willis had resigned as president of the society in 1926 and then drifted away from society activities. By 1928 he had drifted away from earthquake hazard studies and gotten more involved in geological investigations aimed at understanding the structure of the earth's crust. He became president of the Geological Society of America in 1928, and in 1929 embarked on field studies in East Africa.

Although Willis had formally retired from Stanford in 1922 he remained active at Stanford as an emeritus professor until his death in 1949. He thus lived to see a measure of vindication when the magnitude 6.3 Long Beach earthquake struck on the evening of March 10, 1933. Willis did not see the event firsthand, but Los Angeles resident Charles Boudin was among those who did. "I had just left a store in Compton and started across the sidewalk toward my parked car," he

explained, "when suddenly the whole center of the street seemed to
rise. The sudden shock nearly threw me off my feet. Buildings on both
sides of the street collapsed, throwing debris in every direction [fig.
2.6]. Several minutes elapsed before dust from the toppling building
walls cleared sufficiently to see just what damage had been done."
Elsewhere in the area, chemist Robert Green was thrown off his feet,
not by the shock itself but by huge explosions caused by the rupture
of tanks containing sulfuric and caustic acid.

Scientists now know that the 1933 Long Beach earthquake was on
the Newport-Inglewood Fault, one of the faults that Hill had described and dismissed as a harmless relic.

Well-designed buildings fared fairly well in the Long Beach earthquake. Unreinforced masonry buildings did not. A large number of
buildings, including many of the public schools in the area, failed cata-
strophically. In his landmark 1958 textbook, *Elementary Seismology*,
Charles Richter departed from the purely scientific tone of the book
to editorialize briefly. "The calamity," he wrote, "had a number of
good consequences. It put an end to efforts by incompletely informed
or otherwise misguided interests to deny or hush up the existence of
serious earthquake risk in the Los Angeles metropolitan area."

Stark images of collapsed public school buildings accomplished
what all of Willis's energy and passion could not. On March 20, barely
a week after the earthquake, the city of Long Beach voted to adopt the
Santa Barbara building code, which required that all buildings be designed to withstand earthquake shaking. Other cities in the area, including Los Angeles and Pasadena, soon followed suit. State officials
were also, finally, spurred into action. On April 10 the governor signed
into law the Field Act. Although not a building code per se, the Field
Act establishes strict procedures for the design and construction of any
new public school building in California. On May 27 the governor
signed the Riley Act, a statewide law requiring that all new buildings
in California be designed to resist earthquakes. The degree of resistance specified was low—significantly lower in the statewide bill than
in the codes adopted by cities such as Long Beach. By today's stan-

Figure 2.6. Damage to John Muir School in Long Beach caused by the 1933
Long Beach, California, earthquake. (USGS photograph by W. L. Huber.)

dards even the more stringent local codes had not gone far enough.
But these codes, and the Riley Act, were a key first step toward Willis's
vision of a decade earlier.

Willis might have retreated from the scene in the aftermath of his
discredited prediction, but he had not disappeared. In the immediate
aftermath of the Long Beach earthquake he argued that more quakes
were likely and that damaged buildings should be torn down or
strengthened. A March 13 International News Service story out of
Stanford reported that at an earlier meeting of insurance company
presidents in New York City, "Dr. Willis [had] predicted the recent
earthquake." The article quoted Willis directly: "At that time I said the
earthquake might be expected within 3, 7, or 10 years." Willis was
quoted in other articles as saying, "The Inglewood fault which possibly caused [this] disaster is an entirely distinct fault from the San An-

dreas fault." Whether the International News Service story miscon-strued Willis remarks remains unclear. The direct quotes suggest that Willis did not claim to have predicted the quake.

It also remains unclear whether Bailey Willis found himself feeling vindicated or vilified; maybe a measure of both. Without question his frustrations had combined with his passions and impelled him to walk a fine line with statements that, at a minimum, were easily construed as alarmist. By the time Hill had taken Willis's "prediction" apart, the cause of earthquake safety had sustained a serious blow. If not for the 1933 Long Beach earthquake, the blow might have had a more lasting impact. Failed predictions, or the perceptions of such, do garner media attention but do not tend to improve the credibility of earthquake science.

Without question Willis's prediction left a lasting blow to his repu-tation in scientific circles. A scientist today might see it as unfair, that the adjective *discredited* has attached itself to his legacy. But it also serves as a caution, especially insofar as the prediction discredited the larger cause.

As for Willis himself, it might not be clear exactly what he said about the connection between his prediction and 1933 Long Beach earthquake. But in July of that same year, Willis spoke at a meeting of the American Association for the Advancement of Science, and we have a record of those words. "The Wasatch fault," he told the audi-ence in the meeting venue of Salt Lake City, "is a young and active fault. Whether the next quake will come in 10 years or 50 years, no-body knows. But when it does come Salt Lake City should be pre-pared to meet it."

CHAPTER 3

Irregular Clocks

We are predicting another massive earthquake
certainly within the next 30 years, and most likely
in the next decade or so.
—WILLIAM T. PECORA, Testimony to Congress, 1969

Bailey Willis's sense of urgency was based on more than just flawed data. He knew in his bones that California—southern as well as northern—was earthquake country. He knew it had been a long time since the last big quake in southern California. In the 1920s scientists' understanding of faults and earthquakes was still pretty vague, but the idea of an earthquake cycle dates back to the late nineteenth century, when pioneering geologist G. K. Gilbert suggested that earthquakes happened as a consequence of strain accumulation. Once an earthquake released strain, Gilbert reasoned, it would take a certain amount of time for it to build up again.

Gilbert's writings sound almost prophetic when one remembers that at the time, scientists did not yet fully understand the association between faults and earthquakes, let alone what drives the build-up of strain. It would take a few decades for the field to catch up with his vision. We now know that at least some earthquakes occur, if not like clockwork, at least with a certain degree of regularity. We know this because, at least in a broad-brush sense, we understand how earthquakes work. In the mid-1900s earth scientists figured out what drives our planet's earthquake machine. We now know that the earth's thin, rocky crust is broken into a dozen or so large pieces known as plates. Movement of the plates is driven by the slow, steady motion in the earth's mantle.

Some scientific theories are more abstract than others. By the end of the twentieth century, plate tectonics had moved far beyond the realm of abstraction. Using GPS instruments as well as other techniques we can literally watch the plates move. The familiar constellation of continents might look fixed, but it is in fact in constant motion. The motion, although slow, is fast enough and steady enough to literally change the face of the planet while we watch.

At their boundaries, plates generally lock up, unable to move smoothly past their neighbors. The crust thus warps at the boundaries, building up strain, or energy. Eventually this strain releases, abruptly, in an earthquake. The boundaries between tectonic plates are more commonly referred to as faults. Driving into the tiny town of Parkfield, California, on a bridge over a streambed, a small sign welcomes drivers to the North American Plate (fig. 3.1a). The stream looks harmless enough. As visible wracking of the bridge attests, this seemingly small, meandering feature marks the course of the San Andreas, in California the primary plate boundary fault (fig. 3.1b). As such, we expect that, over the long run, this fault will produce more frequent large quakes than any other fault in the state. At Parkfield, as elsewhere in California, at ground level the fault is just one of many landscape features. It is the bird's-eye view that is spectacular. From the air one can see and appreciate the scale of the fault, a sash draped over the state, shoulder to hip (fig. 3.2). The San Andreas Fault is spectacular from the air because it has been hosting big earthquakes for millions of years. When geologists describe faults, they use the term *slip rate*. The term can be confusing to explain because it is not really what it says it is. Except for the short creeping segment in central California, the two sides of the San Andreas Fault do not slip steadily past each other, but remain locked. If you painted a short line across the fault and came back ten

Figure 3.1a. *Top, facing page.* Driving away from the town of Parkfield, California, a sign welcomes drivers to the Pacific Plate. The San Andreas Fault runs along the creek bed beneath the bridge. (Photograph by Susan Hough.)

Figure 3.1b. *Bottom, facing page.* The underside of the Parkfield bridge reveals construction designed to accommodate movement on the fault. (Photograph by Susan Hough.)

Figure 3.2. Seen from the air, the San Andreas Fault cuts a sharp, linear swath running from the bottom right corner of the photograph to the top left corner. The Garlock Fault, which intersects the San Andreas toward the top of the photograph, can also be seen. (LANDSAT satellite view, NASA image.)

years later, most of the time the line would look as straight as the day it was painted. (If the line were long enough it would start to reveal the movement that occurs away from the fault.) Along a locked fault the slip happens only in the abrupt, infrequent lurches. The term *slip rate*, then, refers to the average rate that a fault moves over the long term to keep up with the movement of the plates. Along the San Andreas Fault, this rate is on the order of three to four centimeters per year. If you had painted a line across the fault a million years ago, the two sides would now be separated by about 3 million centimeters, or fifteen miles. About 23 million years ago a block of volcanic rock erupted along the fault, creating new rock on both sides of the fault near present-day Lancaster. One side of this formation, the Neenach Volcanics, remains near Lancaster. The other half, now better known as Pinnacles National Monument, has migrated over two hundred miles north.

In a way earthquake rates are easy to understand, a matter of fourth grade arithmetic. In a large earthquake, for example the one that struck in 1906, a straight line across the San Andreas Fault will be yanked apart by about four to five meters on average. If the slip rate is three to four centimeters per year, simple arithmetic tells us that the fault has to move every 100 to 170 years. This rate can now be confirmed by geological (paleoseismology) studies, whereby geologists dig trenches across active faults and unravel the geological signature of past earthquakes in layers of sediments that have been deposited over millennia. Typically, if a vertical fault cuts through a region where sediments are continually being deposited—for example, a creek bed—then excavation of the site will reveal a layer cake of sediments that have been cut by different earthquakes (fig. 3.3). If carbon is present in the sediments, one can date the ages of the layers with pretty good if not perfect precision. Thus, for example, if one identifies an earthquake break that slices through 1,000-year-old sediments but is capped by unbroken 900-year-old sediments, the earthquake must have happened between AD 1000 and 1100.

Short of a long and detailed historical record, paleoseismology is the best tool scientists have for establishing the rate of past earthquakes on a given fault. But the work is not easy. For starters, a good site is hard to find. The sedimentation rate needs to be fast enough and steady enough that prehistoric earthquakes are reliably trapped between dateable layers, but not so fast as to bury evidence of past earthquakes deeper than can be reached with a backhoe. The work is also painstaking, requiring months if not years of investigation to understand the earthquake record at a single site. But geologists have amassed an impressive body of information from such studies, providing a solid basis for the estimation of long-term earthquake rates on the San Andreas and other faults.

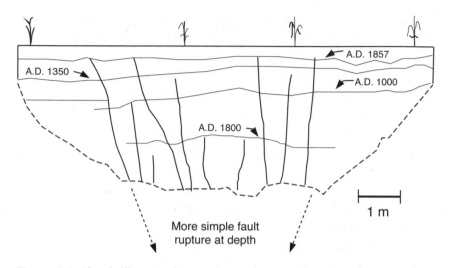

Figure 3.3. Sketch illustrates how sediment layers and earthquake breaks in a trench are identified and labeled. Numbers and shading identifies distinct sediment units. Dark black lines indicate breaks caused by earthquakes. (Image courtesy of Susan Hough.)

These sorts of laws of averages now allow seismologists to forecast the expected rate of earthquakes on a fault such as the San Andreas. As opposed to a prediction of a specific earthquake within a narrow time window, a forecast quantifies, in terms of probabilities, how likely an earthquake is over a period of, typically, decades to centuries. Seismologists also speak of short-term forecasting: the ability to identify earthquakes that are likely to happen on shorter time frames, a few years to perhaps a few decades.

FINE LINES

The distinction between prediction and forecast can start to blur around the edges. If, for example, one identifies a fault that is considered likely to produce an earthquake within the next ten years, this could be considered a prediction or a short-term forecast. Generally speaking, a prediction involves meaningfully narrow specification of time, location, and magnitude.

If the San Andreas has to eventually move three centimeters per year (say) and it has been one hundred years since the last time it moved four meters, and geology tells us that a segment of the fault has big earthquakes on average every 150 years, one might imagine it to be a straightforward matter to calculate the probability that a big quake will happen in the next thirty (say) years. Scientists can and do make these calculations; they are the basis for modern seismic hazard maps. But the simple calculations are not so simple.

WHAT'S SO MAGIC ABOUT 30 YEARS?

When scientists speak to the public and policy makers about earthquake probabilities, we frequently talk about the likelihood of earthquakes in the next thirty years. One might ask, what is so magic about thirty years? The answer is nothing. The tradition dates back to the first "working group" effort, undertaken in the late 1980s and published in 1989, to develop consensus probabilities for future large earthquakes on the San Andreas Fault. The statisticians on the committee wanted to report on the probability of earthquakes in either one-year or one-hundred-year windows. Most members of the committee felt that one year was too short because the probabilities would be too low to get anyone's attention, and one hundred years was too long because the time frame would be too big to be a concern. The committee settled somewhat arbitrarily on thirty years as a span of time that would be within the attention span of the public and decision-makers.

Boiled down to basic arithmetic, a consideration of long-term earthquake rates and past earthquake history on a fault leads to what seismologists call seismic *gap* theory. In particular, if a segment of the San Andreas, or any other fault, has been quiet for a long time, that starts to look like a gap that will eventually be filled by a large earthquake. If it has been a long time since the last big quake on that fault, we start to itching to use the big-O word: *overdue*. It only makes sense. If big quakes happen on average every 150 years and it has been 160 years since the last one, surely it is fair to say a big quake is overdue.

Not quite. An immediate complication arises from the fact that big quakes don't recur like clockwork. Although conceptually it seems straightforward that stress will build and be released in an orderly fashion, it is clear that, at a minimum, the clockwork on any one fault is thrown off by the clockwork on every other fault in the neighborhood.

We can start to understand the irregularity of the clocks not only from geology but also from history, short as it is in California. We know that a big quake happened on the central San Andreas Fault on the morning of January 9, 1857. We also know that a big quake happened, at least to some extent, on an overlapping segment of the San Andreas less than fifty years earlier, on December 8, 1812. Both of these quakes are known not only from geology but also from history. The 1812 quake damaged several of California's missions; the 1857 quake damaged missions as well as some structures in the then small settlement of Los Angeles. It is beyond dispute that these two events struck only a few decades apart. For reasons that we don't entirely understand, big earthquakes on the San Andreas seem to behave like cars on a freeway with light traffic. Sometimes they clump up; sometimes they spread out.

In the mountains north of Los Angeles above the present-day town of Palmdale, the U.S. Army first garrisoned Fort Tejon on August 10, 1854, as part of their efforts to "protect and control" native Chumash tribes. Quartermaster Major J. L. Donaldson chose the site in Cañada de las Uvas, a lovely valley studded with oak trees that had offered a strategic location as well as ample supplies of water. The San Andreas Fault carves an east-west trending valley a few miles south of the site; today Interstate 5 runs along this valley near the small town of Gorman. Here again, viewed from the ground, even with an earth scientist's sensibilities, the fault zone is mostly just another valley. In 1854 nobody understood faults, let alone the detailed geometry of the San Andreas. Had Major Donaldson had a modern understanding of geology, he might have imagined his choice to be a reasonable one. Yes, the site was close to a big fault,

but a big earthquake had struck just a few decades earlier; surely
another one wasn't due any time soon. He would not have imagined
that another big quake, probably even larger than 1812, would
strike just three years later, taking a heavy toll on the fort's masonry
buildings.

In northern California as well, we've seen quakes recur in short
order, geologically speaking, on the San Andreas Fault: the Big One
we all know about, in 1906, and a Big One that we know much less
about, in 1838.

If pairs of earthquakes sometimes pop off fifty years apart on a fault
that has big earthquakes on average every 150 years, it doesn't take
higher math to conclude that the short intervals have to be balanced
by long intervals. Thus, if it has been over 150 years since the last earthquake on a fault that has earthquakes on average every 150 years, that
tells us we could be in the midst of a long interval, and we might have
a while yet to wait.

On a fault like the San Andreas, we do know we won't wait forever.
We don't know how far apart big earthquakes could be on the outside.
Our knowledge of prehistoric earthquakes is derived from carbon-14
dating of sedimentary layers that have been cut by earthquakes. These
dates are always imprecise. Results as well as general geological sensibilities suggest that five hundred years might be a reasonable outside
limit for major earthquakes on the San Andreas, but we really don't
know. Even if we did have good cause to accept this number as a
bound, on a scale of human lifetimes and home mortgages it doesn't
help us much to know that an "overdue" quake could happen tomorrow, or two hundred years from now.

Looking back at the concerns voiced by seismologists over the years
a sense of déjà vu begins to emerge. When Harry Wood first set up
shop in Pasadena in the 1920s he and colleagues were motivated by a
sense that the next great earthquake in southern California might well
happen during their lifetimes. Testifying before Congress in 1969, then
director of the U.S. Geological Survey, William T. Pecora, told a Senate
Appropriations subcommittee that it was inevitable that California

would be hit before the end of the century by an earthquake as large
as the one in 1906. After a magnitude 7.3 quake struck the southern
California desert in 1992, seismologist Allan Lindh, commenting on a
study that estimated a 60 percent chance of a great earthquake on the
southern San Andreas Fault in the next thirty years, told reporters,
"Most of us have an awful feeling that 30 years is wishful thinking."
The sentiments were echoed by seismologist Lucile Jones: "I think
we're closer than 30 years." In January of 2009, a front-page article in
the *Los Angeles Times* informed readers: "The southern stretch of the
San Andreas Fault has had a major temblor about every 137 years, according to new research. The latest looks to be overdue."

The next Big One on the southern San Andreas has clearly been
"overdue" for a very long time. Thus, while today as in Bailey Willis's
time, earthquake scientists know that the southern San Andreas Fault
will produce a large earthquake some day, a sober consideration of history as well as geology tells us that thirty years from now, earthquake
scientists might still be waiting for the Big One to strike southern
California, and sure in their hearts that it will strike in the next thirty
years.

CHAPTER 4

The Hayward Fault

Yesterday morning San Francisco was visited by the most
severe earthquake the city ever experienced. The great
shock commenced at 7:53 a.m. and continued for nearly
one minute, being the largest ever known in this region.
—*San Francisco Morning Call*, October 22, 1868

*T*he southern San Andreas Fault is, of course, scarcely the only fault
we worry about. The earth's crust is riddled with faults, even in regions where earthquakes occur infrequently. But some faults are scarier than other faults. Size matters, but not only that. California's Hayward Fault is clearly a branch off of the main San Andreas trunk, but it is a big branch and, worse, it runs directly through the densely populated East Bay region. The fault runs directly beneath the old City Hall in Fremont, cuts through the Oakland Zoo, and very nearly splits the goalposts at Berkeley's Memorial Stadium (fig. 4.1). Geologists originally tracked the fault as far north as San Pablo Bay, but nowadays are inclined to view the Rodgers Creek Fault as a northward extension of the Hayward (fig. 4.2). The Rodgers Creek Fault winds its way through a less heavily urbanized, and more scenic, region than the Hayward, but it adds an additional measure of concern to the equation. The magnitude of an earthquake depends on the length (also the depth, or width) of the fault that lurches, so a combined rupture of the Rodgers Creek/Hayward faults would be a substantially bigger earthquake than an end-to-end rupture of the Hayward alone.

The Hayward alone is bad enough. This is not a matter of conjecture. At 7:53 on the morning of October 21, 1868, the fault sprang to life. Of the 260,000 people in the Bay Area at the time, most (150,000)

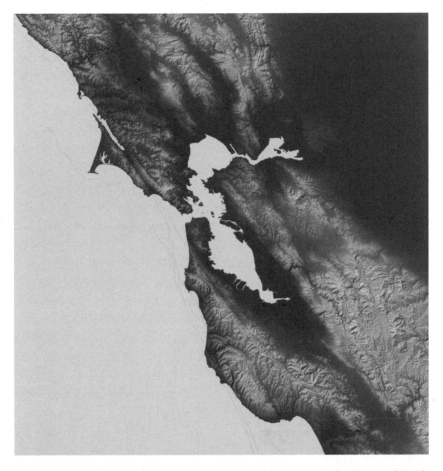

Figure 4.2. The San Francisco Bay Area viewed from space. The Hayward Fault corresponds to the linear feature running through the East Bay Area from the bottom right corner of the image. The San Andreas Fault corresponds to a linear feature running through the San Francisco Peninsula, continuing to the northwest to the north of San Francisco. (Image courtesy of NASA.)

Figure 4.1. *Facing page.* The Hayward Fault runs directly beneath Memorial Stadium, nearly splitting the goal posts, at the University of California at Berkeley. To the right of the archway, steady creep has split the wall enough to create an open crack. (Photograph by Susan Hough.)

lived in San Francisco, but by that time small towns had also sprung up in the East Bay, including Hayward (population 500), San Leandro (400), and San Jose (9,000.) In San Francisco damage was generally concentrated in regions built on soft sediments, or landfill, near the bay. An article in the *San Francisco Morning Call* described the scene: "Yesterday morning San Francisco was visited by the most severe earthquake the city ever experienced. The great shock commenced at 7:53 a.m. and continued nearly one minute, being the longest ever known in this region. The oscillations were from east to west, and were very violent. Men, women, and children rushed into the streets— some in a state of semi-nudity—and all in the wildest state of excitement." Some of city's grandest buildings, including City Hall and the Customs House, were severely damaged.

In the East Bay the earthquake left a thirty-mile swath of destruction along the fault and in areas underlain by soft sediments on the east side of the bay. Almost every building in Hayward was wrecked or seriously damaged, including wood-frame houses that, even without special engineering, tend to withstand shaking fairly well. Closer to the fault many cities in the East Bay experienced even more severe damage. Buildings were also destroyed in San Leandro and San Jose. The jagged surface scar could be traced about twenty miles, from San Leandro to Fremont, suggesting an earthquake of approximately magnitude 6.8. But earthquakes are known to keep going deep in the crust beyond the extent of the surface break, and a 1996 study by Paul Segall and Ellen Yu concluded that the 1868 quake had done just that. Looking carefully at early surveying data, Segall and Yu concluded that the quake had continued northward as far as Berkeley. The larger extent (about thirty-two miles) implies a larger magnitude, closer to 7 than 6.8.

The difference between 6.8 and 7 might not sound like a big deal, but by virtue of the logarithmic nature of the magnitude scale, a 0.2 unit increase in magnitude corresponds to about twice as much energy release. Worse, because shaking is especially severe in proximity to the fault, a longer break puts more real estate directly in harm's way. Finally, prior to Segall and Yu's study scientists had generally believed

that the Hayward Fault was split into a northern and a southern segment, each of which would break independently. If the 1868 quake
had extended as far north as Berkeley then the two segments had broken together. Here again one arrives at the conclusion that the fault
could be a bigger menace than previously believed.

But how frequently will damaging earthquakes like the 1868 event
occur? U.S. Geological Survey geologist Jim Lienkaemper has spent
most of his career working to answer this question. He has not only
mapped the fault in fine detail; he has also spent most of his career literally in the trenches, painstakingly piecing together a chronology of
prehistoric earthquakes on the fault.

Thanks to the exhaustive efforts of Lienkaemper and his colleagues,
the Hayward Fault has given up over a millennium of geological secrets. By 2007 Lienkaemper and his colleague Pat Williams were able
to present a 1,650-year chronology of earthquakes on the fault, identifying twelve different events including the 1868 quake. Anyone who
has paid attention to a Bay Area newspaper in recent years is probably
familiar with the oft-touted result that the Hayward Fault has produce
large (magnitude ≈ 7) quakes every 140 years, on average. With the last
quake in 1868 this is a result that hits you where you live, if you live
along or anywhere near the fault. With such a long record of such apparently regular quakes, surely the next one is due any day now.

The next Hayward Fault quake could indeed strike at any time; it
could strike before this book is published. No question, anyone who
lives anywhere near the San Francisco Bay Area needs to be concerned
about earthquakes. But should one feel concerned, or should one feel
panicked?

The year 2008 sounds like an auspicious date if earthquakes occur
every 140 years on average and the last one was in 1868. In fact, however, 140 years is the average time between the last five Hayward Fault
earthquakes. Stepping back to consider the longer record, it appears
that the Hayward Fault has been a bit of an overachiever in recent
centuries, producing more large quakes than one expects on average
over the long run. This could mean that the rate of earthquakes on the

fault is actually increasing, but much more likely reflects nothing more than the usual vagaries of irregular geological clocks.

Considering the full record, the average interval between successive large quakes is close to 160 years. The median time between quakes, meanwhile, is close to 150 years—which is to say, looking at the time between pairs of successive quakes, half of the times are less than 150 years and half are longer. That the median differs from the average tells us that half of successive events are closer together than 150 years, but a small number of pairs are separated by significantly more time, and the handful of long times pushes the average up. By way of analogy, imagine a neighborhood in which most of the houses sell for less than $150,000, but with a small number of houses that sell for much more. The average would be skewed by a small number of data points, while the median would be a better reflection of the prices of most of the houses in the neighborhood.

Since earthquakes have apparently occurred more frequently than normal over the past seven hundred years we could be in the midst of a cluster of events that will continue for at least one more quake. But the other possibility is that the fault experienced a temporary flurry of quakes in the past seven hundred years and so the next quake might be slow in coming, perhaps in part because the 1906 quake on the San Andreas cleared a lot of stress from the system. Also, it appears likely that the northern and the southern segments of the fault can go off independently, producing quakes closer to M6.7 to 6.8. If the 1868 quake was indeed longer than some, it might have cleared out the system, so to speak, such that it will take longer than average for stresses to build enough to cause another quake. A further consideration is that at the present time the fault is experiencing a fairly high rate of steady motion, or creep, along much of its extent (fig. 4.3). We don't fully understand how the fault works, in particular if it always or only sometimes experiences creep. It is possible, at least, that fault creep waxes and wanes over time, and that the ongoing process of creep serves to slow the build-up of stress that will drive earthquakes.

Looking at the record of past earthquakes one can use various simple approaches to forecast the time of the next quake. Drawing a line

Figure 4.3. At Nyland Ranch in San Juan Bautista, steady creep on the San Andreas Fault has caused an offset in the wood rail fence. (Photograph by Susan Hough.)

through just the last five events, one of course predicts the next quake in 2008. If one bases a forecast on a line drawn through all twelve points, the result jumps all the way out to 2095, 227 years after 1868. But no matter how one slices the data, the probability of a big quake remains too high for comfort.

At a minimum there are good reasons to not be panicked that 2008 has already come and gone. And so, here again, even with a relatively regular irregular clock, one arrives at conclusions that seem as frustrating as they are useful. From a societal hazard mitigation point of view, the high probability of a quake in thirty years tells us most of what we need to know. Any house, commercial building, or piece of infrastructure in the East Bay area is very likely, albeit not certain, to be shaken severely during its lifetime. Enforcement of building codes, retrofitting of vulnerable structures, public education and awareness campaigns— these are priorities that scientists and many agencies are taking seriously. The scientists who study the fault and the hazard it poses are doing their best to get the message out. "We are trying to get people to do the right thing," says USGS seismologist Thomas Brocher. "An earthquake of this size is larger than most cities can handle."

Indeed, from the point of view of an individual in a craftsman cottage in Berkeley, or an apartment building in Hayward, a high thirty-year probably tells you most of what you need to know, but none of what you *want* to know. Is the next quake going to be in 2010 or 2039? Could it be as late as the closing years of the twenty-first century? Here again, we do know it will strike one day. For his part, Jim Lienkaemper, the unassuming—and not so terribly old—geologist who has arguably had more intimate contact with the Hayward Fault than any other scientist, has been heard to quip that he is not sure it will happen during his lifetime.

CHAPTER 5

Predicting the Unpredictable

> "There's no such thing as earthquake weather," he said.
> "Eeyore was just being gloomy again."
> "Oh, bother!" said Pooh, who had missed his lunch
> because of Eeyore.
>
> *—Eeyore, Be Happy!*

*I*f the long-term rate of earthquakes on a given fault isn't regular enough to tell us when a big earthquake is due then our ability to make meaningful short-term forecasts is limited at best. Bailey Willis and Harry Wood had a sense that it had been a long time—maybe too long—since the last big quake in southern California. Eighty years later scientists still have that sense. Earthquake scientists have felt the same sense of urgency about the same fault for eight decades; this alone tells us something.

But there is earthquake forecasting and there is earthquake prediction. Even if earthquakes were completely irregular in their occurrence, we could still predict specific earthquakes if the earth gives us some sort of heads up that a large earthquake is on the way. Earthquake prediction has been called the Holy Grail of seismology, but really, *prediction* is the goal; earthquake *precursors*—signals from the earth indicating a large earthquake is coming—are the grail. There is reason to believe that the grail is not entirely a creature of myth and legend.

Some earthquakes are in fact predictable. After an earthquake happens its aftershocks are quite predictable. We can't predict individual aftershocks any more than we can predict individual earthquakes, but fairly simple rules, based on countless observed sequences, tell us how many aftershocks of various magnitudes we can expect.

Scientists have also identified repeating earthquakes, very small earthquakes that recur regularly along segments of faults that do not lock up, but are able to move via steady creep, including the ninety-mile stretch of the San Andreas Fault in central California. Along this and other creeping faults, small (magnitude 2–3) quakes are generated by small, isolated patches along the fault that do get stuck, and then break. In these locations, it seems, the earthquake cycle (stress accumulation, stress release) is not buffeted by cycles on nearby faults, and is thus able to proceed on a more regular timetable. In recent years scientists have identified a number of sequences of repeating earthquakes, and have been successful in predicting successive events within a time frame of months. A team of Japanese scientists led by Naoki Uchida has identified a repeating sequence of more healthy quakes: a series of impressively regular magnitude 4.8–4.9 events that has struck offshore of northeast Japan.

In recent years scientists have had some success predicting magnitude 5–6 quakes along certain kinds of faults—transform faults—along mid-ocean ridges. In this setting, recent results show that faults start to move slowly, in so-called slow slip events, before moderate earthquakes. They also tend to be preceded by foreshocks that follow identifiable patterns.

We've also had a fair measure of success in predicting the earthquakes that humans manage to trigger, in particular by injecting fluids into the crust. Any man-made disruption of fluids will perturb the crust in ways that can cause earthquakes to happen. When a significant earthquake strikes close to an area of active oil production—for example, a magnitude 4.3 event in south-central Texas in 1993—scientists and the public sometimes suspect that the quake might have been induced by hydrocarbon extraction or injection of fluids. A cause-and-effect relationship is usually impossible to prove. But where fluids are pumped back into the crust, for example at geothermal plants where brine is reinjected, the process clearly raises fluid pressures in the crust, causing the rock to fracture. These earthquakes are small as a rule, but longtime residents of the Clear Lake region in northern California are convinced that the increased rate of magnitude 4+ shocks in recent decades can be traced directly to deep reservoir drilling that began in the early 1970s. Scientific analysis tends to support this conclusion. Careful analysis also supports the conclusion that fluid injection at the Rocky Mountain Arsenal outside of Denver was responsible for inducing a magnitude 5.3 earthquake that caused $1 million in property damage in 1967. We believe, for what it's worth, that disruptions associated with fluid injection and/or extraction will are not substantial enough to generate large earthquakes. We could be wrong about this. Earthquakes can also be triggered when large reservoirs are impounded, and we know these earthquakes can be significant events. One such induced quake, the 1967 magnitude 6.3 Koyna event in central India, caused serious damage and nearly two hundred deaths.

In any case, if scientists can predict magnitude 3 quakes on the creeping section of the San Andreas Fault or magnitude 5 quakes along mid-ocean ridges, the results are scientifically interesting but societally unimportant. The grail that we care about—the grail that remains elusive—is the identification of reliable precursors prior to damaging earthquakes. Many earthquakes are preceded by foreshocks, smaller quakes that strike close to the epicenter of the eventual main shock. The problem is, when a small earthquake strikes there is no way to tell a foreshock from a garden-variety earthquake until a bigger quake happens. To be useful for prediction, a precursor needs to, first, occur reliably before most or many large earthquakes, and, second, occur only before a large earthquake. Even researchers have been known to overlook a key point about earthquake precursors. It is possible for a documented precursor to be real—that is, to be associated with physical processes that presaged a later earthquake—but to be of no practical use for prediction because the precursor does not reliably presage large earthquakes.

Returning to the tiny town of Parkfield, tantalizing anecdotal evidence suggests that the 1966 Parkfield earthquake was preceded by precursors. Weeks before the earthquake, Clarence Allen led a field trip to Parkfield with a group of visiting Japanese scientists. During this excursion, the scientists noticed what appeared to be fresh cracks along the fault. Then, at some time during the twenty-four hours prior to

the June 28 event, a pipe crossing the fault broke. Although no monitoring instrumentation was in place to record exactly what the fault
was doing in the weeks and hours prior to the earthquake, one had to
think that it must have started to move slowly, creating fresh cracks
and eventually breaking the pipe.

The tiny town of Parkfield looms large in the annals of earthquake
prediction. The town itself is bucolic, nestled in a central California
corridor that is more remote than one might imagine any place in
California could be. Here, a century and a half after the Ranchero era,
cattle ranching remains an active way of life. The Parkfield Cafe (fig.
5.1) serves up offerings ranging from the Big One (a one-pound steak)
to a selection of Aftershocks (desserts), but it's a thirty-minute drive to
the nearest gas station or grocery store. As early as 1906, in the aftermath of the San Francisco earthquake, geologists realized that this
stretch of the fault had been visited by especially frequent moderate
and large earthquakes. It is unclear if geologists deduced this from the
landscape or by talking with the locals. In any case, by the 1980s seismologists with the U.S. Geological Survey had taken a formal tally of
known past quakes and concluded that magnitude 6 quakes had struck
Parkfield on average every twenty-two years. In the late 1980s they
stepped forward with a bold announcement, that the next "Parkfield
earthquake" would strike within four years of 1988.

The U.S. Geological Survey came to town and set up shop along
with other agencies, including the California Geological Survey. The
area was blanketed with monitoring instruments, not just seismometers but other types of instrumentation as well. When the next Parkfield earthquake struck, these instruments would be waiting. The year
1988 came and went. Then 1998. Some instrumentation fell by the
wayside but the commitment to monitoring remained, not only with
a measure of growing unease but also with a continued conviction
that Parkfield was still the best place in California if not the world to
hope to record a moderate quake. Not to mention the fact that earthquake monitoring instrumentation has tremendous inertia. It takes a
lot of time, effort, and resources to install a seismometer or other type
of monitoring instrument. Apart from the logistics of the installation,
one must first find a suitable site; then one has to get permission from

Figure 5.1. The Parkfield Café invites guests to "Be here when it happens!" (Photograph by Susan Hough.)

a landowner. If scientists head out to the field imagining the permitting process to be a simple task, they quickly learn the error of their ways. Permitting can easily delay a project, sometimes kill it altogether. Once an individual or an organization succeeds in getting an instrument installed, nobody has the heart to pick it up again.

But sometimes it takes a lot of heart to keep instruments going. For over thirty years scientists at the U.S. Geological Survey in Menlo Park operated an array of creep meters, instruments that would record any small motions such as those that apparently preceded the 1966 earthquake. Colleagues credit John Langbein for heroic efforts to continue operating the instruments as the years ticked by after 1992, and many started to doubt that the earthquake would ever happen. Other instruments at Parkfield, including strong-motion seismometers deployed by USGS seismologist Roger Borcherdt and his team, were also kept alive more by sheer individual determination than by organizational commitment.

In the 1980s geophysicist Roger Bilham installed four creep meters across the fault at Parkfield, focusing his own monitoring efforts along the part of the fault where the pipe had broken. His funding ran out in the late 1990s, and the instruments fell into a state of disrepair. By the beginning of 2004 Bilham had secured funding to bring his creep meters back to life. On September 27, 2004, he visited his instruments in Parkfield to reprogram them to cut the recording rate in half, recording a data point every minute instead of every thirty seconds, so that he wouldn't need to visit the instruments as often. The change was safe enough, he remembers thinking as he drove away from Parkfield, the silly earthquake would probably never happen.

Rarely is one proven wrong quite so quickly. At 10:15 in the morning, local time, on September 28, 2004, Langbein, Borcherdt, Bilham, and the rest of the earthquake science community got the earthquake they had been waiting for. It was thirty-seven, not twenty-two, years after the last one, and, unlike the 1966 Parkfield earthquake the fault broke from south to north rather than north to south. But otherwise, in terms of magnitude and extent of fault break, it was the earthquake that scientists had expected. Most scientists now regard Parkfield as a qualified success, notwithstanding the fact that the 2004 quake was "late" and "backward." And most scientists now regard this section of the fault as a good illustration of the irregular regularity that characterizes earthquakes. In this one location, scientists had been able to make a meaningful forecast on a time scale of decades. The payoff in this case was not hazard mitigation: the sturdy wood-frame structures in the Parkfield area rode out the shaking quite well. The payoff was science. The bounty of data from the Parkfield earthquake has given seismologists a number of important new insights into how faults and earthquakes behave, and the nature of earthquake shaking.

But if the Parkfield earthquake was discouraging in its implications for meaningful short-term forecasting, it was downright depressing in its implications for prediction.

The real point of the Parkfield Prediction experiment had been to set a trap. The next time the fault broke in a moderate quake, sophisticated instruments would be in place to record exactly what had happened along the fault before the earthquake. And so they did. Prior to the 2004 earthquake, the scores of instruments at Parkfield—the seismometers, strain meters, magnetometers, etc.—recorded nothing out of the ordinary. Bilham's creepmeters had been resuscitated just in time to show that the fault did not start to creep in advance of the earthquake. The crust did not start to warp; no unusual magnetic signals were recorded. The earthquake wasn't even preceded by the large foreshock scientists were expecting based on the fact that the 1966 and 1934 Parkfield earthquakes were both preceded by magnitude 5ish foreshocks approximately seventeen minutes earlier. In 2004, no foreshock activity was recorded in the hours/days prior to the earthquake.

The San Andreas Fault did experience increased creep in 2004, not at Parkfield but about 125 kilometers to the north in the town of San Juan Bautista. Movement along this part of the fault caused a pipe to break three days before the 2004 Parkfield earthquake. According to our understanding of faults and earthquake nucleation, the small motion at San Juan Bautista cannot have exerted any significant influence on the fault at Parkfield. Thus the coincidence, while intriguing, is thought to be just that: coincidence.

Considering the negative results from Parkfield, some seismologists were quick to speak of nails and coffins. Time to write the eulogies: earthquake prediction was finally, officially dead. Many seismologists regarded this pronouncement as premature, but few seismologists were inclined toward optimism. On the one hand, the negative results from Parkfield do not prove that the earthquake was not preceded by precursors, only that this particular earthquake was not preceded by any precursors large enough to be detected on instruments. On the other hand, if precursors can't be detected with the density of instruments that we had in place in Parkfield, what hope do we have for detecting them anywhere?

The negative results from Parkfield do not prove that significant precursors don't precede some earthquakes. But they do prove that significant precursors don't precede all earthquakes.

The quest to find precursors will not be easily deterred. The knowledge that, at any moment, the world around us could become unglued, shaking the structures around us to rubble, is not an easy awareness to live with. We want, desperately, to believe that, if earthquake prediction isn't possible right this second, it will be possible, soon.

And, negative results from Parkfield notwithstanding, there are scientific reasons to believe that prediction might be possible. A big earthquake is, after all, a portentous event. Over one thousand kilometers of plate boundary lurched during the 2004 magnitude 9.3 Sumatra earthquake. Waves from the earthquake literally rang the earth like a bell. During these vibrations, every point on the earth's surface was set into motion by at least one centimeter (approximately 1/3 inch). These distant undulations were not felt because the ground moved so slowly, each wave taking on the order of a half hour to pass from peak to trough. Even for those of use who imagine we're familiar with earthquakes, it can be hard to get one's head around the kind of energies that are required to make an entire planet dance.

Surely, one imagines, a large earthquake doesn't just happen. If stresses build up for hundreds or thousands of years, surely, or so one imagines, the earth sends out some sorts of signals as a fault reaches the breaking point. Laboratory studies show that rocks subjected to increasing stress start to crack in predictable ways before they break. Alternatively, perhaps large earthquakes finally occur because of some sort of external trigger that we could identify. Both sorts of ideas have been explored by serious scientists. The notion of precursory warping, or strain, remains on the table. Studies of old triangulation data suggest that the seafloor beneath Tokyo Bay might have experienced anomalous warping in the years prior to the great 1923 Kanto earthquake, which devastated Yokohama and Tokyo. And in the hours before the magnitude 8.1 Tonankai earthquake offshore of southwest Japan, instruments recorded a large apparent tilting of the seafloor.

Other anomalies have been reported in advance of other earthquakes. In 1997 Max Wyss and David Booth tallied a list of documented precursors that were credible and apparently significant. Reviewing observations that had been touted in the scientific literature as precursors, they considered whether proposed precursors had been clearly defined and shown to be statistically significant. In the end they were left with only five observations that belonged on a preliminary list of significant earthquake precursors. These included a gradual water level rise over three days preceding a magnitude 6.1 earthquake in central California, groundwater changes before a magnitude 7.0 earthquake in Japan, and several documented changes in local earthquake patterns.

In general, possible earthquake precursors fall into several broad categories: (1) hydrological or hydrogeochemical changes, for example changes in well or stream level, or in the chemistry of groundwater; (2) electromagnetic signals, for example anomalous electrical currents or ultra-low-frequency magnetic signals, and so-called earthquake lights; (3) changes in the physical properties of the earth's crust, for example changes in the speed of seismic waves; (4) changes in the patterns of small or moderate earthquakes; (5) anomalous warping of the crust; and (6) anomalous release of gas or heat along a fault.

Earthquake precursors also potentially play out over different time scales. Some apparently anomalous signals have been recorded seconds-to-hours before large earthquakes, others days-to-weeks earlier.

Within the seismological community the quest to find precursors has focused largely on earthquake patterns. When scientists first installed seismometers in southern California they were motivated in part by the hope that small quakes would herald big quakes. As Charles Richter wrote in a memo in 1952, "There was original hope that little shocks would cluster along the active faults and perhaps increase in frequency as a sign of the wrath to come." By 1952, Richter and his colleagues had been disabused of this notion. "Too bad," he wrote, "but they don't. Roughly, little shocks on little faults, all over the map, any time; big shocks on big faults."

Richter's pessimistic pronouncement notwithstanding, the line of research has continued. Dr. Win Inouye, a Hawaiian-born Japanese seismologist, concluded in 1965 that small quakes become less frequent prior to large earthquakes. A number of more recent studies have also identified purported patterns of precursory quiescence. Pioneering Japanese seismologist Kiyoo Mogi described another precursory pattern, what scientists know as the Mogi doughnut: a decrease in activity near the site of a future earthquake combined with an increase in activity in a doughnut-shaped region around it. To date the existence of Mogi doughnuts have not been proven by rigorous statistics. Seismologists tend to have their doubts about doughnuts, and quiescence. But in the back of our minds, we wonder.

Earthquake precursors are a matter of not only science but also lore and legend. So deeply rooted is the notion of "earthquake weather" that it figures prominently in the plot of *Eeyore, Be Happy!* first published in 1991 as part of the Little Golden Book series for children. The belief also continues to run deep that animals can somehow sense impending large earthquakes. What the animals might be responding to, if anything, remains unclear. But if animals are able to sense and respond to something, then animal behavior itself could be considered a precursor. The roots of this idea are older than earthquake science. In recent decades scientists have undertaken experiments to test the idea rigorously. All of these attempts have failed. For example, researchers have compared the number of lost pet ads in newspapers to the record of significant earthquakes, reasoning that if animals do sense impending earthquakes they would likely become agitated and more likely than usual to run away from home. Such studies have revealed that, not surprisingly, there is a correlation between lost pet ads and major storms; that is, animals don't sense impending storms but rather run away during them. Such studies reveal no correlation between the number of ads and earthquakes. In the 1970s seismologist Ruth Simon devised experiments to see if cockroach activity could be correlated to impending earthquakes. It couldn't. During the 1970s the U.S. Geological Survey undertook or sponsored several other studies, including one to look systematically at the behavior of rodents in the southern California desert. None of these investigations ever bore fruit.

Nor did Swedish researchers have better luck when they placed earthquake sensors—call them seismoometers—on the backs of a cows in southern Sweden. When a magnitude 4.7 quake struck just a few miles from the herd, the animals not only failed to show any unusual behavior prior to the quake, they also failed to react *to* the quake. "One can probably say," observed researcher Anders Herlin, "that, as a species, cows are not the world's most earthquake-sensitive animals."

And yet the lore lives on. As recently as 1976, not quite the ancient past, editorials and articles in major newspapers described anomalous animal behavior as established fact. Of the many possible reasons behind the enduring longevity of the belief is the fact that not even seismologists can fully explain what snakes, frogs, and other animals were up to in northern China during the winter of 1974/75.

CHAPTER 6

The Road to Haicheng

Although the prediction of the Haicheng earthquake was
a blend of confusion, empirical analysis, intuitive
judgment, and good luck, it was an attempt to predict a
major earthquake that for the first time did not end up
with practical failure.
—KELIN WANG, QI-FU CHEN, SHIHONG SUN,
and ANDONG WANG, 2006

*E*verybody knows that animals provide clues that big earthquakes
are coming—wasn't unusual animal behavior the basis for that
earthquake that was predicted in China?

The magnitude 7.5 Haicheng earthquake, which struck northern
China in 1975, looms large in earthquake prediction annals. To understand the political as well as the scientific importance of the earthquake it is useful to first retrace the steps along the road to Haicheng.
If earthquake prediction represents an ongoing collision between science and society, that collision surely reached a crescendo during the
years immediately preceding and following Haicheng.

Within earth science circles the middle of the twentieth century
was largely defined by the plate tectonics revolution. For seismologists
this meant understanding earthquakes in a global context; at long last
understanding the forces that cause earthquakes to happen. Yet seismology remained a cottage industry, with even the best earthquake
monitoring networks left to run on a shoestring. During the Cold
War–era seismology suddenly found itself center stage, a key player in
matters of national strategic importance. Almost as soon as nuclear
testing began, scientists as well as officials recognized the unique role

that seismology could play, providing recordings of waves generated by large underground explosions. Governments might keep their nuclear tests secret, but waves from blasts respect neither international borders nor government classifications. By the late 1940s the government had begun to fund seismologists, offering contracts to scientists in academia and launching their own internal programs.

To monitor nuclear explosions, in particular outside the United States, one needs a global network of seismometers with readily accessible data. The United States could launch classified programs, but by the 1950s the government had put substantial weight and its resources behind the establishment of the World-Wide Standardized Seismograph Network (WWSSN), the first global standardized, modern network of instruments. Military patronage provided resources on a scale previously unimaginable to seismologists. Not only that, but resources for research also poured into the field under the auspices of the Vela Uniform program. The origins of the name *Vela* are somewhat obscure; according to some it was derived from the imperative form of Spanish verb *velar* meaning "to watch," or "to keep vigil over."

To some extent the field of seismology was inevitably shaped by the interests of its benefactors. But decisions made early on provided that, first, the data would be freely available, and, second, that a sizable fraction of the new research money would support basic rather than targeted research. The WWSSN and overall Vela program thus proved to be a tremendous boon for seismologists. With a newfound bounty of global data, many—although not all—top seismologists, understandably, focused on global studies.

On May 22, 1960, a massive earthquake struck Chile, unzipping almost a thousand kilometers of the offshore plate boundary—with an estimated magnitude of 9.5, the largest earthquake seismologists had ever recorded. The death toll in Chile was low (1,655) for an earthquake of this magnitude. But the earthquake and subsequent tsunami left some 2 million people homeless. And as the 2004 Sumatra earthquake and tsunami illustrated, a large tsunami can have a long reach. Thus tsunami waves from the great Chilean earthquake not only crashed onto local shores but also sped westward across the Pacific

Figure 6.1. Damage to Hilo, Hawaii, from a tsunami generated by the 1960 Chilean earthquake. The run-up at Hilo was measured at 10.7 meters (35 feet). (Photograph courtesy of National Geophysical Data Center.)

Ocean, in the open ocean moving quietly at the approximate speed of a modern jet airplane. With no warning systems in place the wave approached Hilo, Hawaii unheralded (fig. 6.1). In the open ocean a tsunami wave involves up-and-down movement of the entire water column; because so much water is moving, the waves generate only gentle swells at the surface (fig. 6.2). As the waves approach coastlines, all of the energy gets concentrated in progressively smaller volumes of water. Wave heights grow.

Some shorelines around the world turn out to be natural tsunami catchers. Hilo is one of them. On May 23, 1960, nearly fifteen hours after the Chilean earthquake struck, ten-meter (thirty-five-foot) waves crashed into Hilo Bay. Sixty-one people were killed; wood-frame buildings were crushed or swept away. Even on the far side of the Pa-

Figure 6.2. On April 1, 1946, a large earthquake in Alaska generated a strong tsunami that propagated across the Pacific Ocean. The waves reached Hilo, Hawaii, approximately 5 hours later, producing a surge as high as two 3-story buildings and devastating the waterfront. People along the shore ran for their lives. The calamity claimed 159 lives in Hawaii and caused $26 million dollars in damage. (Photograph courtesy of National Geophysical Data Center.)

cific Ocean the waves remained powerful enough to take a toll, causing 138 deaths in Japan and thirty-two in the Philippines.

Seismologists and engineers around the world rushed to study different aspects of the Chilean earthquake. Presented with unprecedented data from seismometers around the planet, American seismologists focused again on global studies. For the first time, they had compelling observations of the earth's so-called free oscillations. The earthquake had been large enough to set the entire planet ringing. Observations of this ringing, as well as other types of waves, gave seismologists new tools with which to probe the inner structure of the planet.

For the field of seismology, these were exciting times. Seismologists might not be ghouls, but the fact remains, our most exciting data is often somebody else's nightmare.

Some seismologists were particularly mindful of this. In the early 1960s, Lou Pakiser headed a group of seismologists at the U.S. Geological Survey office in Denver. His group received Vela funding for investigations of the structure of the crust, a natural direction for the USGS given their original mandate of mapping mineral resources. Like his colleagues in academia, Pakiser and his group had some latitude in their Vela-supported investigations. In the early 1960s, in the aftermath of the Chilean quake, the group's research interests began to migrate from studies of the earth's crust to studies of earthquakes.

The ball had begun to roll. On March 27, 1964, it picked up speed when another massive earthquake struck, this time along the coast of Alaska. The great Good Friday earthquake also did not cause huge loss of life, but 131 people died, and dramatic images of damage in Anchorage from the earthquake and tsunami were beamed around the world, a Technicolor disaster for the early television era. In U.S. government as well as scientific circles earthquake monitoring and hazard assessment were suddenly of concern. Whereas the new WWSSN stations provided good recordings of large earthquakes at large distances, these instruments get blown off scale by large earthquakes at close distances, and were not sensitive enough to record small earthquakes at close distances. Different instruments are required, respectively, strong-motion and weak-motion instruments. As of 1964, strong-motion monitoring in the United States was limited; the Coast and Geodetic Survey operated a total of about three hundred instruments. Weak-motion, or regional seismology, had largely been left to academic institutions such as Caltech, U.C. Berkeley, and St. Louis University.

A panel of experts led by Frank Press from MIT was convened in the aftermath of the Alaska quake. By the fall of 1965 they made a series of urgent recommendations to the White House, in particular a ten-year, $137 million program aimed at predicting earthquakes and mitigating earthquake damage. USGS leadership recognized the opportunity that had bitten them in the butt. Just days after the panel delivered their recommendations the USGS came forward with a bold announcement: a new research center was being established in Menlo Park, California. As reported by cub reporter David Perlman, the new center would be "designed to assault the barrier that has long prevented earthquake prediction." Capitalizing on earlier momentum, Pakiser led the charge, leaving Denver, where he headed the Center for Crustal Studies, and moving to Menlo Park, California, to run the new National Earthquake Research Center.

It would take another decade before the government actually funded the ambitious new research program that had been recommended. In the meantime, resources were brought to bear on the problem. In the late 1960s the California Division of Mines and Geology, which had formerly focused on mapping mineral resources, stepped up their funding for earthquake research. The USGS also began to support its nascent program. This led, in the words of historian Carl-Henry Geschwind, to politically embarrassing "interagency squabbling" between the USGS and the Coast and Geodetic Survey, which had been responsible for strong motion recording since 1932. By some accounts the Coast and Geodetic Survey was elbowed out of the way, having, in Geschwind's words, "acquired a reputation for engaging only in routine data collection rather than innovative research." By other accounts CGS's home agency, the National Oceanic and Atmospheric Administration, decided to jettison their earthquake program when budget cuts threatened to jeopardize higher-priority programs.

While these machinations were playing out in a political arena, a long period of seismic quiet in the Los Angeles area ended abruptly, one minute after six o'clock on the morning of February 9, 1971, when the magnitude 6.6 Sylmar earthquake struck the northern San Fernando region north of Los Angeles. A Veteran's Administration Hospital in Sylmar collapsed; a community hospital in the nearby community of Olive View sustained severe damage (figs. 6.3a and 6.3b). Sixty-five lives were lost, and property damage totaled half a billion dollars.

In recent decades seismologists have learned that an earthquake like Sylmar will be followed by not only aftershocks close by but also by

Figure 6.3a. Damage to the Olive View Hospital in San Fernando from the 1971 Sylmar, California earthquake. (USGS photograph.)

more distant triggered earthquakes. It is also clear that a noteworthy earthquake like Sylmar triggers not just earthquakes, but earthquake predictions. When a large earthquake garners media attention, the pseudo-science community springs into action with predictions of future events and, often, claims to have predicted the recent event. In December of 1971, self-proclaimed "earthquake prophet" Reuben Greenspan predicted that a devastating earthquake would strike San Francisco on January 4, 1972.

But the pseudo-science community is by no means alone. An ardent critic of earthquake prediction claims throughout his lifetime, Charles Richter spoke out with signature bluntness in the aftermath of the Sylmar quake. People who claim to make precise predictions were, in his words, "charlatans, fakes, or liars," an epithet he later and famously streamlined to "fools and charlatans." While no scientists disputed Richter's assessment of current prediction abilities, many stepped forward to voice optimism about the likelihood of prediction

Figure 6.3b. The Sylmar earthquake generated a surface offset, or scarp, the result of vertical movement of a thrust fault that continued to the surface. The feature has eroded significantly since 1971, but this small segment has been landscaped and incorporated into the design of a fast-food parking lot. (Photograph by Susan Hough.)

in the not-too-distant future. In February of 1971, Caltech Seismo Lab head Don Anderson said that, with suitable funds and research, "it would in my opinion be possible to forecast a quake in a given area within a week." In June of 1971, geology professor Richard Berry was quoted in the *Oakland Tribune* as saying, "If you had asked me about the possibility of earthquake prediction 10 years ago I probably would have said no. Now it looks like that within a decade we will have information that, within limits, will make it possible to predict what will happen."

Where had this optimism come from? Some of it can be traced to the former Soviet Union. By the early 1970s word began to spread in the U.S. earthquake science community that scientists in the Soviet Union had had success with prediction using a method to detect subtle changes in earthquake wave speeds in the crust. In particular, Soviet scientists began to look for changes in the ratio between the speed of P (pressure) waves and the speed of S (shear) waves in the crust, what scientists know as the Vp/Vs method. The method, which ultimately derived support from sound theories, has a measure of conceptual appeal: if earthquakes are preceded by the formation of cracks in the earth, such changes are expected to change the ratio of wave speeds. And the ratio P-wave to S-wave speed can be measured more precisely than the wave speeds themselves.

Interest in earthquake studies and earthquake prediction in the Soviet Union traced back to a series of three large quakes that had struck Soviet Central Asia, near Garm, Tajikistan, between 1946 and 1949, including a magnitude 7.4 event on July 10, 1949. By 1970 Soviet seismologists were reporting apparently impressive progress with the Vp/Vs method. In 1971 the International Union of Geodesy and Geophysics (IUGG) meeting was held in Moscow, and Soviet seismologists generated a buzz with what appeared to be exciting results. News of the apparent Soviet success made its way to the West. The day after a magnitude 5.8 earthquake rattled the southern California coast near Ventura in February of 1973, readers of the *Los Angeles Times* were informed that "no one doubts the validity of the Soviet research" on the Vp/Vs method.

Peter Molnar, then a post-doc at Scripps Institution of Oceanography, was able to join a field trip to Garm that was planned in conjunction with the 1971 meeting. Molnar describes the experience as a "fun trip," but one that did not allow participants to dig into the details of the Soviet research program. Intrigued, he returned home and later applied for a National Academy of Sciences fellowship to spend time in the Soviet Union. Specifically he proposed visiting the seismological institute in Garm to learn more about the Vp/Vs method. Landing in Moscow in 1973 he found hosts who were not initially keen to arrange his trip to Garm. While he was there, a high-level, VIP delegation of earth scientists, including Frank Press, did visit Garm, and apparently greased the skids: Molnar was soon on his way to Central Asia.

The head of the institute, renowned Soviet scientist Igor Nersesov, was away when Molnar arrived, but two young scientists, Vitaly Khalturin and his wife Tatyana Rautian, took Molnar under their wing and made sure he had access to key seismograms. Molnar set out to make measurements from the seismograms, in particular the precise times of P and S waves for each earthquake at each station. With this information, he could reproduce the calculations that Soviet scientists had done to calculate Vp/Vs values before and after significant earthquakes. Three weeks later Molnar found himself unable to reproduce the intriguing results he had seen. His calculations of Vp/Vs revealed a lot of scatter, but no discernible patterns prior to large earthquakes.

Nersesov returned to the institute and took Molnar through his own analysis of the same data. Molnar quickly realized how the game was played. Researchers had produced maps of how Vp/Vs varied across their study region, including systematic differences related to rock type as well as splotchy regions where Vp/Vs was locally higher or lower than average. These results were in keeping with expectations: seismic wave speeds, and the ratio of P- to S-wave velocities, vary somewhat throughout the earth's crust. When Nersesov calculated an anomalous Vp/Vs value from a particular seismogram, he had wiggle room in the interpretation; he could attach the value to a previously identified anomalous patch in the crust, or he could declare it to be a temporal anomaly.

By way of analogy, imagine using a series of CAT scans to decide if a box of raisin bran is being invaded by mealworms. A baseline scan would reveal the locations of the raisins, with some inevitable fuzziness. Over time subsequent scans would reveal small anomalous signals. But any one apparent anomaly might just be part of a previously identified raisin. It would be a matter of judgment, deciding which anomalies were real (i.e., which were mealworms), and which could be discounted (i.e., which were only parts of raisins). If one made one's interpretations blindly, without any outside information, one would be left with considerable uncertainty but no real bias. A flurry of anomalous blips would provide compelling evidence that an invasion was underway.

Nersesov's interpretations, however, were not blind. He knew what seismograms he was looking at, and he knew the times of the large earthquakes that had struck within the region. Molnar understood that Nersesov and his colleagues had not been consciously dishonest in their interpretations. They had, however, set up the rules of the game such that it was all too easy to fool themselves.

Upon his return to the United States Molnar explained his findings to colleagues, including two of the scientists who had been part of the VIP visit to Garm: Frank Press and Lynn Sykes. Molnar encountered skepticism from the two men, both of whom remained convinced that the anomalies were real.

Top brass at the USGS also remained eager to learn more. In 1974 they dispatched one of their young scientists, Rob Wesson, to Garm. Wesson describes the experience as "culturally more than scientifically educational." He arrived in Garm to find posters on the wall, similar if not identical to the summary presentations at the IUGG meeting. But the seismologists responsible for the studies were nowhere in sight. In fact nobody at the institute was prepared to discuss the work in detail. Wesson left Garm with the impression that Soviet seismologists in fact had something less than full confidence in the results that had been touted at the 1971 meeting.

Wesson's impressions were later confirmed to Peter Molnar. Although Molnar had quickly backed away from prediction research he maintained a close professional and personal relationship with his colleagues Rautian and Khalturin. Many years later they told Molnar that they had viewed his visit as a sort of a test: would the young American scientist buy into what they recognized as wishful thinking, or would he arrive at the answer they themselves knew to be honest?

If the Garm pathway failed to live up to expectations, some in the United States remained undeterred. In 1973, a team of researchers at Lamont-Doherty Geological Observatory used the Vp/Vs method to make an apparently successful prediction of a small (magnitude 2.6) earthquake in upstate New York. Also in 1973, a team of Lamont scientists led by Christopher Scholz developed the theory of dilatancy, the process whereby the volume of rocks increases slightly (dilates) when subjected to stress. The increase occurs because rock grains are shifted, generating new cracks. The theory, which appeared to provide a sound physical basis for the promising observational results, generated excitement among Scholz's colleagues at Lamont and elsewhere. The link between deep fluid injection and earthquakes, first established in the 1960s and developed through the 1970s, also gave scientists a better understanding of the role that fluid pressures can play in earthquake nucleation. For the first time, it seemed, scientists were getting a handle on exactly how and why earthquakes happen.

Within the ranks of the mainstream earthquake science community momentum started to build for an ambitious new federal National Earthquake Hazard Reduction Program to support earthquake science. Whereas top scientific leadership had earlier capitalized on a symbiosis with national strategic interests to boost funding for seismology, in the 1970s hazard reduction had come to the fore as an issue of national concern. As causes go it isn't a bad one. Seismology is a relevant science by virtue of the contributions it can make to seismic hazard assessment, and risk mitigation. Reliable estimations of ground motions from future earthquakes represent the basis of effective building codes, to cite just one example. Even if we could predict earthquakes with perfect accuracy, major earthquakes would be catastrophic economic disasters if buildings and infrastructure don't survive the shaking.

The problem with earthquake hazard assessment is that, as causes go, it just isn't sexy. It isn't what the public wants, and one can't dazzle Congress with promises of better building codes. Better building codes tend to be a bitter pill: they jack up construction costs for a payoff that might or might not ever come. Bold promises of earthquake prediction, especially in the wake of an earthquake like Sylmar, *was* sexy stuff. By the early 1970s ambitious plans began to take shape, and gain momentum, for a new national program. And these plans were promoted largely if not entirely on promises of earthquake prediction— promises that were, in retrospect, somewhere between optimistic and totally irresponsible.

If the Sylmar earthquake brought the prediction pot to a simmer, the spark that sent it into a full, rolling boil came from half a world away. In terms of intrigue, one could scarcely have scripted a better plot twist. In early 1975 news began to reach western shores: seismologists in China had apparently successfully predicted the massive magnitude 7.3 Haicheng earthquake, preventing the enormous loss of life that would have surely happened had the earthquake struck unheralded. In 1975 China was, of course, just beginning to emerge from the throes of the Cultural Revolution. Mao was ill but alive, and China remained very much closed off from—and enigmatic to—the outside world.

That the earthquake happened was beyond dispute. By 1975, waves from an earthquake this large were recorded on a worldwide network of seismometers, providing seismologists around the world with sufficient data to determine location and magnitude. But what had happened in China prior to the earthquake, this was less clear. The party line was clear, spoken with conviction and repeated many times over. The line is parroted to this day by authoritative sources within and outside of China, "The Haicheng earthquake was successfully predicted, saving untold thousands of lives."

It would take decades for western seismologists to understand that this statement resides somewhere in the murky realm between fact and fiction.

Seven months after the earthquake, Robin Adams from New Zealand was the first non-Chinese seismologist allowed to visit the Haicheng area. In China he encountered a sea of blue Mao jackets, one obvious indication that the obsession with the Cultural Revolution remained very much alive even as its leader's health was failing. Seismologists told Adams, "Chairman Mao told us to predict earthquakes, so we had to." Pressing repeatedly, Adams learned that several predictions had led to evacuations prior to 1975 and had not been followed by earthquakes. Adams's hosts were vague on this point. Asking about the details of the prediction prior to the Haicheng earthquake, the script was far clearer: "Seismology walks on two legs, the experts and the broad masses of the people." The "broad masses" were encouraged to be part of the effort, making simple observations of water level and animal behavior. To Adams's mind, this was "an elaborate social bluff to make people believe that they were involved, and cleverly in the case of failure to divert the people's criticism away from the experts themselves."

And yet in 1975 a prediction had, it seemed, been made, and an earthquake had happened, and lives had been saved. The world took note. Scientific leaders in the West took note. Earthquake prediction, it seemed, was indeed possible. The full story of the prediction would not be told until records were made available decades later. A 2006 article in the *Bulletin of the Seismological Society of America* by Kelin Wang and his colleagues provided a story that western seismologists had only known in part. Digging back to original documents, Wang's team was able to piece together the full story. The bottom line: yes, the earthquake had been predicted, sort of. Whether this one qualified success held any general promise for reliable prediction was another story.

Geologically speaking the road to Haicheng began with a series of moderate earthquakes in the late 1960s and early 1970s in the normally quiet Bohai Sea region southwest of Haicheng and east-south-east of Beijing. As seismologists have come to appreciate in recent decades, earthquakes do cluster in time and space. A large earthquake disturbs its surroundings in a way that makes earthquakes on neighboring faults more likely. Adjacent dominoes can topple minutes, days, or years after being disturbed—or not at all (fig. 6.4). By the late 1960s, Chinese seismologists had already developed more vague but similar theories for how large systems of faults could become activated.

A secret 1970 report from provincial authorities stated, "Epicenters of recent strong earthquakes in the Bohai Bay area show a tendency of northward migration (fig. 6.5). Jinxian and Yingkou that are located on the Bohai Bay may fall into this area of strong earthquakes and suffer destruction." Premier Zhou Enlai, a strong proponent of earthquake prediction, called a national meeting on the subject that same year. A year later, the State Seismological Bureau (SSB)—roughly speaking, China's equivalent of the U.S. Geological Survey earthquake program—was created. In 1972 the SSB began to organize regular conferences on earthquake prediction.

As official prediction efforts gained steam officials launched a broad

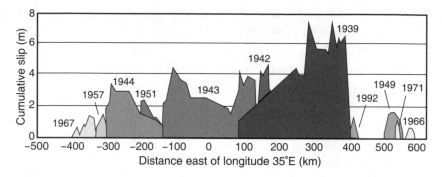

Figure 6.4. Progression of earthquakes on the North Anatolian Fault in northern Turkey. A large earthquake in 1939 increased stress at both ends of the rupture, apparently triggering a cascade of subsequent large events that played out over the rest of the twentieth century. The progression was noted prior to 1999, when two large (M7.4 and M7.2) earthquakes struck on the North Anatolian Fault to the immediate west of the sequence. (Adapted from work by Ross Stein, Tom Parsons, and others.)

awareness campaign to educate the public. This effort included basic earthquake education as well as so-called citizen science. Pictorial brochures, for example, showed individuals how they could make amateur observations of well water levels, animal behavior, even electrical currents in the earth.

In June of 1974 the SSB organized a conference focused specifically on assessment of future earthquakes in the Bohai Bay region. A consensus report stated that "most people think that earthquakes of magnitude 5–6 may occur in this or next year in Beijing-Tianjin area, northern Bohai Sea [region.]" One notes that this statement stopped well short of sounding alarm bells regarding the prospects for future large earthquakes. The situation was discussed at some length. Seismologists at the conference fell into two camps: those who argued that earthquakes in the region were uncommon, so future events were not to be expected any time soon; and those who argued that there was imminent danger of magnitude 7 quakes.

Monitoring continued in the region. Apparent anomalies cropped up in tide gauge records and in geomagnetic recordings; both were

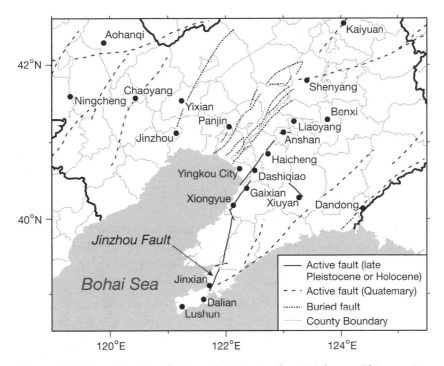

Figure 6.5. Map showing faults and cities in the Haicheng, China region. Epicenter of the 1975 Haicheng earthquake is indicated by the star. (From Wang et al., *Bulletin of the Seismological Society of America*, June 2006, reprinted with permission.)

soon recognized to be spurious. In 1974 leveling studies revealed an apparent warping of the crust. It remains unclear if this observation was caused by groundwater extraction or if it indeed was related to the impending earthquake. At the time, in any case, it fueled concerns.

With marching orders to predict earthquakes Chinese seismologists met in June of 1974 to consider the potential for future large earthquakes in north-northeast China. This group identified six places—five of them far away from Haicheng—that could be at risk of significant earthquakes within the next few years. Seismologists at the conference were far from agreement on most key points, including not only the likely location and magnitude of a future large earthquake but also theories underlying prediction.

In December of 1974 a moderate, magnitude 5.2 quake struck the northern Bohai Sea region, once again in a region that had experienced few earthquakes previously. A team of researchers concluded that the filling of a nearby reservoir had likely caused this quake. The team's report did not reach provincial officials until late January; in the meantime concerns took a quantum jump upward. The provincial Earthquake Office held an emergency overnight meeting on the event of December 28. The following day they announced that an earthquake around magnitude 5 might still occur in the Lioyang-Benxi region. Two days later, they specifically predicted that an earthquake was likely to occur by January 5.

Rattled (so to speak) by recent earthquake activity, provincial officials took the prediction seriously. Military leaders, including General Li Boqui, came forward with bold statements. "This time," he said, "we will demonstrate 'preparedness averts peril.' It's like fighting a war. Be prepared for a big war, early war, nuclear war, and sudden attack." Officials were well aware of the potential for false alarms; they regarded them as the price one had to pay for security.

January 5 came and went without further notable activity. The false alarm did have consequences. Reportedly, six hundred workers at a petroleum production field in Panjin, citing a fear of future earthquakes, left their posts to return home for extended New Year's vacations.

By January 10 the provincial Earthquake Office announced that the likelihood of additional significant quakes in the Liaoyang-Benxi region was low. By the end of the month authorities had received the report on the December 22 earthquake sequence.

But as natural earthquake activity apparently subsided, amateur observations of groundwater fluctuations and strange animal behavior continued. Records indicate that snakes and frogs reportedly came out of the ground, something snakes and frogs are not supposed to do during winter hibernation season. With no notable further earthquake activity through January these observations did not generate a sense of urgency among officials who were possibly by that time weary of urgency.

On February 1 and 2 several small earthquakes were detected near

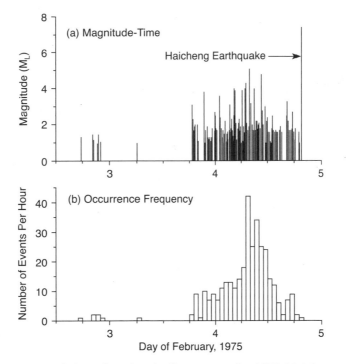

Figure 6.6. Recorded earthquakes leading up to the 1975 Haicheng earthquake. Bottom panel shows number of recorded events per hour. Top panel shows magnitude of events. (From Wang et al., *Bulletin of the Seismological Society of America*, June 2006, reprinted with permission.)

Haicheng—here again, the activity was in the same general region as previous events, but no quakes had been recorded previously in the immediate area. This sprinkling of quakes by itself did not engender concern. On the evening of February 3 the Haicheng region began to rumble in earnest. By midnight the nearby Shipengyu observatory had recorded thirty-three earthquakes. The quakes continued through the wee hours of the morning, intensifying in frequency and magnitude by daylight. A magnitude 5.1 quake struck at 7:51 a.m.

By midday on the 4th a number of other moderate shocks had struck (fig. 6.6). The provincial Earthquake Office made a report to provincial officials at two o'clock, describing the activity and the damage caused by the moderate quakes. Shaking had been severe enough to topple chimneys and even damage gable ends of masonry buildings that were not designed for earthquake resistance. By the evening of the 4th earthquake activity had abated somewhat, although the local observatory continued to record small shocks and several events were large enough to be felt.

Concern might have abated as well, but it didn't. The remarkable burst of events set balls in motion at various levels, in various directions, and with various degrees of grounding in legitimate science.

At the grassroots level, the spate of felt and even damaging shocks was enough to make a strong impression on people who had been steeped in the rhetoric of earthquake preparedness and prediction, and citizen science.

In Yingkou County, adjacent to Haicheng County, Cao Xianqing had served as head of the local (county) Earthquake Office since the office was created in 1974. A former army officer, Cao had taken up his new duties with great enthusiasm. He supervised a network of amateur observatories in his county. Although he was accused at least once of causing panic with incendiary rhetoric, he brought great energy to the job of collecting amateur observations, including underground (telluric) current readings and water well levels. He had seen continuing anomalies in December and January, and was convinced they were harbingers of an impending large quake. Under his direction, by late January all of the communes in the county had opened their own earthquake offices.

When the ground began to rumble on the evening of February 3, Cao sprang into action, spearheading an emergency meeting of party officials in the county at 8:15 the next morning. Cao informed the committee that "a large earthquake may occur today during the day or in the evening." He asked the County Committee to "please take measures." They did. Their edict went out immediately after the meeting. All meetings were to be canceled, all public entertainment and sporting events suspended, all business activity suspended, all production work halted. Communes were further to be instructed to make sure local families left their houses.

The word went out, by messenger as well as loudspeaker. Many in the Yingkou County countryside slept outside that night. In the throes of a typically cold north China winter it was not a decision made lightly.

Neighboring Haicheng County had no Mr. Cao, but they did have a team of amateur observers at the Haicheng Observatory who made telluric current observations during the unrest of February 3–4. Around dawn, a couple of hours before the 7:51 a.m. shock, their readings showed a large increase. The readings increased again just before 2:00 p.m., leading the team to warn that a moderate or larger earthquake might occur within a few hours. The statement was delivered by bicycle messenger to the County Earthquake Office. News also spread rapidly by word of mouth and telephone as the observatory team as well as the messenger informed other individuals.

The sense of urgency in Haicheng County lagged behind that in Yingkou. The county government did hold a meeting on February 4, but not until 6 p.m. The citizenry of Haicheng was generally aware that a large earthquake might strike, but their concern did not rise to the level that convinces people to sleep outside in subfreezing temperatures. Some in the area, having heard reports or rumors of possible quakes, did go to bed—or put their children to bed—wearing winter clothing. Sporadic evacuations did happen that night in the county, based on decisions of local commune officials and individuals. But unlike in Yingkou County evacuations were not extensive.

The Haicheng County Earthquake Meeting wrapped up after 7:00 p.m. with no official pronouncement. Through late afternoon and evening only a handful of small earthquakes were recorded, probably none big enough to be felt.

At 7:36 p.m. a previously unknown fault near the boundary of Yingkou and Haicheng counties suddenly sprang to life, unleashing the fury of a magnitude 7.3 earthquake on the surrounding countryside. The ground itself was shaken strongly enough to give way in many places. In areas underlain by artificial fill, ground slumped and shifted, underground pipelines ruptured. In other areas, natural sand layers liquefied, creating dramatic sand blows. Even well-built masonry buildings collapsed in the shaking. In Haicheng forty-four people died in the collapse of a three-story guesthouse. The overall death toll of the earthquake was relatively low: around two thousand from both direct and indirect causes such as fires and hypothermia. A number of babies were brought safely to shelters only to die from suffocation, the result of desperate attempts by caregivers to keep them warm.

In the aftermath of the earthquake Peking Radio reported that Chairman Mao Tse-tung and other leaders were "greatly concerned" about loss of life and property. Although the initial reports did not provide details, China specialists in the West expressed concern that the fact that it was mentioned at all indicated that the disaster must have been of major proportions. By March 14 a few more details made their way to western journalists. An individual in China wrote in a letter to a relative in Hong Kong, "the earth cracked, people could hardly stand, and many chimneys and old houses collapsed." By this time the official Chinese news agency, Hsinhua, was reporting that seismologists had been able to provide advance warnings that helped prevent losses.

The low number of fatalities cannot, however, be explained entirely by the evacuations. Even in regions where evacuations were sporadic or nonexistent the death toll was relatively low for such a large earthquake in an urban region. Researchers now point to three explanations for this. First, houses in the region were traditionally built with a wood frame and brick in-filled walls. The first layer of the roof was usually wood, covered by either bricks or straw. A small wood-frame house, even of simple design and construction, has a natural resistance to shaking. Although brick walls fell during the Haicheng earthquake, many walls remained standing, preventing catastrophic collapse (fig. 6.7).

A second explanation for the low death toll was that, at 7:36 in the evening, people were at home, not at work. The most dangerous buildings in the region were the modern structures built entirely of unreinforced masonry: office buildings, factories, guesthouses, schools. Many of these suffered heavy damage, just as happened three decades later when a magnitude 7.9 quake struck the Sichuan Province on May 12, 2008. Whereas the 2008 quake struck in the afternoon when children were in school, schools and other large structures were empty, or nearly empty, when the Haicheng earthquake struck.

Figure 6.7. Damage from the 1975 Haicheng, China, earthquake. The number of fatalities was low in part because of the prediction, but also in part because of the prevalence of wood-frame houses in the area. Such structures, even when heavily damaged, tend not to collapse completely. (Courtesy of Kelin Wang.)

The third fortuitous happenstance also resulted from the time of the quake. In the early evening, some people—presumably many children—were asleep, but many were still awake, and had their wits about them. People were able to leave their homes after the shaking began, and to snatch sleeping children from their beds.

As urban earthquake disasters go, the ending of the Haicheng story was better than most. And clearly it helped that the earthquake was, at least to some extent, anticipated. Having reviewed the full story as pieced together in the 2006 article by Kelin Wang, Qi-Fu Chen, Shihong Sun, and Andong Wang, one can revisit the party line: "The Haicheng earthquake was successfully predicted, saving untold thousands of lives." The sleuthing efforts by Wang and his colleagues reveal that there was no official prediction by top officials at the national level. The upsurge of activity in northern China was noted and discussed in the years preceding the earthquake, but the meaningful assessments and announcements in the final days and hours came from the county and grassroots level. That county offices existed in the first place can be credited to actions taken at higher levels. Similarly, the campaign of awareness and citizen science had instilled a broad awareness of earthquakes—whether or not people understood the issues fully.

Of the overall saga, the actions taken by Cao Xianqing stand out as singularly decisive and effective in saving lives in Yingkou County. Records show that throughout the day on February 4, 1975, Cao continued to urge immediate evacuations. His sense of urgency was largely based on gut-level concern about the escalating foreshock sequence. But his statements were remarkably precise. He stated that the earthquake would strike before eight o'clock that night, and that the magnitude would depend on the exact time: magnitude 7 at seven o'clock, magnitude 8 at eight o'clock. This patently absurd aspect of the forecast was apparently based on some sort of extrapolation of foreshock activity. The earthquake struck at 7:36 p.m. and registered magnitude 7.3; the seismologist can offer no explanation beyond the often correct but always unsatisfying answer, coincidence.

But why was Cao so sure that the quake would strike by eight o'clock? In interviews following the earthquake, he explained that he was motivated by something he had read in a book, *Serendipitous Historical Records of Yingchuan*. According to this book, heavy autumn rains would "surely be followed" by a winter earthquake. Cao explained that rains had been very heavy in the fall of 1974. The escalating unrest thus clearly pointed to a large winter earthquake, and according to the

Chinese calendar, winter ended at 8 p.m. on February 4. Thus the certainty, the earthquake had to happen by this time. In fact Mr. Cao's calculations were off by an hour: winter officially ended at 6:59 p.m. that night. The earthquake was actually a half-hour late.

There are times when one is right for the wrong reasons. Then there are times when one is right for spectacularly wrong reasons.

Cao's intuition about energetic foreshock sequences was, however, perhaps on the money, and supported by at least a measure of scientific evidence. Seismologists have long known that about half of all earthquakes are preceded by foreshocks, smaller earthquakes close to the location of a subsequent larger "mainshock." For reasons we still don't understand, some large earthquakes have no foreshocks, some have a few foreshocks, and some have a lot of foreshocks. The Haicheng earthquake had a lot of foreshocks.

Foreshocks are unique among proposed earthquake precursors in that we know they are real. But for prediction purposes they are mostly if not entirely useless. The problem is that, as noted, while half of all earthquakes are preceded by smaller foreshocks, there is nothing about a foreshock that distinguishes it from the great multitude of small shocks that are not followed by anything bigger. The only glimmer of hope that foreshocks might be identified as harbingers of subsequent large earthquakes is some evidence that energetic sequences of foreshocks, when they occur, are more tightly clustered in space and time than garden-variety small quakes. This remains controversial today, and may well be wrong. But the final sequence preceding Haicheng fit the pattern.

The regional unrest in the years preceding Haicheng also fit a pattern that at least some seismologists believe is real. In addition to traditional foreshocks, some regions appear to become active over a few years or decades. We don't entirely understand why this happens. As noted, some theories point to stress changes, a mechanical toppling of dominoes. Or perhaps the more plastic deep layers, below where earthquakes happen, are set into slow motion after big earthquakes, eventually pushing nearby faults over the edge. In any case, Chinese seismologists were ahead of their western colleagues in their recognition of regional earthquake clustering, and this factored into the Haicheng story.

And then, as earthquake prediction aficionados like to remind us, there were those other anomalies. Currents in the earth, water levels, elevation changes, frogs and snakes.

Underground current measurements are notoriously noisy. If one is comparing a constantly fluctuating measurement with the timing of earthquakes, it is not hard to find apparent precursory signals. Mountains of measurements made by amateur observers, including school-children, posed something of a challenge for scientists. Although provincial officials often discussed these measurements at meetings, they were apparently not taken terribly seriously, as they were never quoted in official reports.

Reported anomalies in water wells and animal behavior are seemingly more compelling. Winters are cold in northern China; it is a rash and probably deadly move for a snake to emerge from hibernation. The ground was not exactly crawling with snakes, but there were almost one hundred documented snake sightings in the Haicheng region in the month prior to the earthquake.

How does one account for this? The simplest explanation is that they were disturbed by foreshock activity, including shocks too small for humans to feel. Or perhaps the reported sightings reflected only a sort of mass hysteria on the part of those doing the sighting. There is no question that the mind is capable of extraordinary invention, if suitably primed. But neither explanation sits entirely well. The numbers of reports—frog/snake sightings and well-level fluctuations—did not track either the earthquake activity or the sense of urgency among local people and officials. Either explanation could be right; the truth is, we don't know. Haicheng leaves us with hints and perhaps promises, but in the end far more questions than answers.

Before exploring the intriguing possibilities for what the Haicheng observations could mean, the Haicheng story inevitably closes on a cautionary note. Whatever happened, if anything, in the Haicheng region to disturb frogs and snakes and water wells, one thing is clear: it does not happen commonly before large earthquakes. While anecdotal accounts of odd animal behavior invariably surface after large earthquakes, rarely do they come close to the scale of reported anomalies in the Haicheng region. Kelin Wang and his colleagues suggested that perhaps reptiles in more seismically active regions, like California and Japan, do not react to impending earthquakes because earthquakes are common, whereas in north China they are rare. This explanation rings false on biological grounds: animals cannot develop instincts regarding something their ancestors never, or only very rarely, experienced.

WHEN SNAKES AWAKE

In 1978 Helmut Tributsch, a professor of chemistry at the Free University of Berlin, the book *Wenn die Schlangen Erwachen*. The English translation, *When the Snakes Awake*, was published in 1982 by the Massachusetts Institute of Technology press. The book is an interesting amalgamation of anecdotal accounts accepted uncritically but also serious science. Tributsch advances the hypothesis that charged aerosol particles—bits of matter as small as a few molecules to as large as one micrometer—are responsible for anomalous animal behavior prior to large earthquakes. That charged aerosol particles could effect animals, as well as produce other electromagnetic effects, is not an outlandish idea. The tripping point is that electrical fields would have to be generated in the earth to release charged particles into the atmosphere, and, while Tributsch advances several theories, the arguments here are less compelling.

If Haicheng's snakes and frogs were reacting to something, that something was, if not entirely unique, then at least unusual. In the wee hours of the morning of July 28, 1976, a magnitude 7.5 quake struck the city of Tangshan, about 120 kilometers east of Beijing. Amateur prediction efforts were alive and well in the Tangshan region, as well as other parts of China, before the earthquake. In the words of Kelin Wang, "some appeared to have captured some interesting anomalies." But, he continues, the attentions of the State Seismological Bureau were not focused on this region, and the observations from Tangshan were not closely scrutinized or considered prior to the earthquake. "This resulted," Wang observes, "in a large and confusing volume of prediction records for Tangshan. The situation makes it easy … to select for materials to make sensational stories but makes it very difficult for researchers to sort out." Purported anomalies included strange lights the night before the quake and disturbances in water well levels, as well as other geophysical signals. Wang concludes that observations of earthquake lights—reportedly seen just moments before the shaking was felt—are credible. As for the rest of the purported precursors, whether they were real or not, they were not sufficient to focus the attentions of seismologists or government officials on the Tangshan region.

Whereas Haicheng is a story of good fortune at almost every turn, the stars aligned differently, metaphorically speaking, for Tangshan. About a million people lived in the industrial city, many in masonry buildings with very poor earthquake resistance (fig. 6.8). The quake struck without warning, and in the dead of night. At 3:42 in the morning almost everyone was sound asleep when the ground began to shake and homes began to crumble. The residents had no time to react, no chance to escape. Thousands must have died instantly or nearly so, crushed by debris or suffocated. The powerful quake left the city without power; those who lay injured in the rubble were left without assistance until daybreak. Eighty kilometers (fifty miles) away in Tientsin (now Tianjin), the shock was powerful enough to knock over heavy furniture and literally tear apart the Tientsin Friendly Guest House, a newly built structure.

By official government estimates, about 250,000 people lost their lives in the Tanghan quake. The number is horrific, but many believe the true toll to be at least twice—maybe as much as three times—as high. Whatever had been brewing, so to speak, beneath the earth's surface to generate the foreshocks and other strange signals before Haicheng had not been repeated here. If we someday understand fully the launch sequence that preceded Haicheng we might have the basis for prediction of some earthquakes. The *if* and *might* in this sentence loom

Figure 6.8. Aerial view of widespread damage caused by the 1976 Tangshan, China, earthquake. (USGS photograph.)

large; so does the *some*. Some respected seismologists argue that earthquakes will never be predictable. It is also possible that some earthquakes might be more predictable than others. The prediction of some earthquakes would certainly be a good thing, but it would not be an entirely satisfying solution. We would still be left with the knowledge that a large earthquake could strike anywhere, any time, with no warning whatsoever.

CHAPTER 7

Percolation

Definitive and consistent evidence for hydrological and
hydrogeochemical precursors . . . has remained elusive.

—MICHAEL MANGA and C-Y. WANG

*L*ooking back at the events leading up to the Haicheng earthquake,
one is left—at least, some are left—with the sense that something
must have been brewing under and around the Bohai Sea in the years
leading up to 1975. It bears repeating: the obvious inference might
well be wrong. Earthquakes may cluster in time and space for no other
reason than the fact that movement on one fault disturbs neighboring
faults. It would not be good news for the cause of prediction, if earthquakes do cluster only because of so-called earthquake-earthquake interactions. If successful earthquakes in a region do essentially topple
one another, we can look at fallen dominoes and understand how the
toppling played out. But whether one domino will topple another
might well depend on details that we can't hope to know or predict.
At present, at least, there doesn't seem to be any way to tell in advance
when a neighboring fault will be disturbed to the point of producing
another big earthquake. But the alternative, namely that clusters of
events like the 1960s–1970s quakes in the Bohai Sea region are driven
by some sort of underlying disturbance, holds somewhat more hope. If
we could recognize the underlying disturbance—again, assuming it
exists—we might know that earthquakes are on the way.

What of this possibility? Could the subsurface percolate before
large earthquakes?

In a few cases the answer is clearly yes. In the later summer of 1965
a remarkable earthquake swarm began in the Matsushiro basin, in Na-

gano City in central Japan. The swarm, which continued for two years, produced thousands of quakes. At its peak the swarm included about one hundred earthquakes a day that were strong enough to be felt. During the swarm large volumes of ground water laden with gases including carbon dioxide were released; still today the area is riddled with active springs and high CO_2 concentrations. Careful analysis of the gases points to a source aquifer fifteen kilometers deep in the crust. A team of Japanese scientists led by Norio Yoshida concluded that an impermeable sheet of rock, which had formerly kept the aquifer safely confined, broke in 1965, allowing large volumes of pressurized water to rise into the crust, causing fracturing and weakening of the rock. The team further found water with a similar chemical signature in another region where earthquake swarm activity had occurred.

There is evidence that fluids play a role in other earthquake sequences, including swarms of small earthquakes in volcanic and geothermal regions; also the earthquakes that have occasionally popped off throughout historical and recent times near the small town of Moodus, Connecticut.

In few cases is the link between hydrology and seismic activity so clear. But the idea that groundwater has something to do with earthquake processes is scarcely outlandish. The ground beneath our feet might look solid, but even at great depth we know there is more to the underworld than layer upon layer of intact rock. Subsurface strata are especially lively in subduction zones and volcanic regions. Along the former, where slabs of sea floor sink down to great depths, minerals are subjected to progressively higher pressures and temperatures, and undergo what geologists know as phase transitions. In short, minerals get squeezed and turn into other minerals. This process creates by-products, including water. Being more buoyant than its surroundings, water rises away from the sinking slab, making its way up along—and creating—cracks. As the water encounters rocky material in the overlying crust, it allows some material to start melting at lower temperatures than melting can otherwise begin. Thus, while some textbook illustrations might suggest otherwise, magma is not actually created along the sinking sea floor, but in the overlying crust.

In recent years, seismologists have observed a new kind of pulse from the earth, known as nonvolcanic tremor. (The phenomenon of volcanic tremor results from the movement of magma through the subsurface plumbing beneath active volcanoes.) Along some subduction zones, including those in the Pacific Northwest and Japan, a low hum is sometimes generated as the seafloor sinks beneath the overriding plate. In Japan, careful analysis of these signals by a team of U.S. and Japanese researchers points to locations at the interface between the slab and the overriding plate. Although the results to date do not prove conclusively how the signals are generated, most studies have concluded that they are facilitated if not generated by fluids along the plate interface. One working hypothesis is that fluids essentially lubricate the plate boundary, allowing it to slide, generating chatter along the way, once stresses build to a certain level.

In any case we know that fluids are generated along subduction zones, and beneath chains of subduction zone volcanoes. Subduction zones generate most of the volcanoes we see on dry land; the remaining 5 percent are hot-spot volcanoes like the Hawaiian Islands, thought to be the result of a long-lived upwelling, or plume, of magma from deep in the earth's mantle. (Most of the planet's volcanic activity occurs underwater, at the mid-ocean ridges where new oceanic crust is generated.) Clearly lively fluid systems exist beneath all types of volcanoes. In any deep volcanic system gases are produced as magma melts and interacts with the surrounding rock. The most common of these gases are water vapor, sulfur dioxide, and carbon dioxide. Volcanoes can release small amounts of other gases as well, including hydrogen sulfide, hydrogen, carbon monoxide, hydrogen chloride, hydrogen fluoride, and helium.

Volcanic eruptions can release huge volumes of gases into the atmosphere. Volcanoes can also sometimes burp quietly, releasing gases without an eruption. In May of 1980 the Long Valley, California, volcanic system came to life, producing several moderate earthquakes as magma moved upward beneath the central caldera. The unrest has continued in on–again, off–again fashion. The latter half of 1989 was one of the on–again times; a swarm of small earthquakes occurred be-

Figure 7.1. Trees killed by carbon monoxide near Mammoth Lakes, California. (Photograph by Susan Hough.)

neath Mammoth Mountain, and again monitoring instruments detected an upward migration of magma deep in the earth.

The 1989 episode again ended without a bang. But the following year Forest Service rangers noticed areas of dead and dying trees on one side of the mountain. Tracts of trees can die off as the result of drought or infestation, but these causes were investigated and eliminated. USGS scientists confirmed what Forest Service rangers suspected: the trees were being killed by high concentrations of carbon dioxide in the soil (fig. 7.1). Carbon dioxide is heavier than air; eventually it dissipates once it reaches the surface, but high concentrations can build up in soils, and in surface depressions. The gas isn't just lethal to trees. In 2006 three members of the Mammoth Mountain Ski Patrol died after falling into a carbon dioxide–filled snow cave created by a gas vent.

In active volcanic regions, earthquakes are driven by volcanic processes. For reasons we don't entirely understand, magma blobs are only sometimes on the move. As magma makes its way up to the surface, the overriding crust crackles and pops in response. New fractures are created, earthquakes occur, and gases are sometimes released. If magma rises far enough, the volcano erupts. When this happens, earthquake activity does not taper off but rather builds in a way that scientists have learned to recognize. Thus volcanic eruptions are, for the most part, predictable.

For the most part. Our crystal balls for volcanoes are better than for earthquakes, but they are not perfect. A large, complicated system like Long Valley can start to percolate and then stop. And then start and then stop. In late 1997, the percolation showed signs of building toward a runaway process, with thousands of earthquakes recorded in the weeks prior to Thanksgiving. Facing the decision to raise the alert level from "green" to "yellow," seismologist David Hill found himself feeling "chartreuse." Just as chartreuse edged toward yellow the earthquake sequence showed indications of slowing, and the USGS did not issue a prediction that would have turned out to be a false alarm. But they came close.

Other predictions have been issued over the years, a few false alarms but more than a few successes, including the 1980 prediction of the Mount St. Helens eruption.

But what about the lion's share of earthquakes that are not associated with active volcanic processes? So far as we understand—or at least, so far as scientists now generally believe—earthquakes occur in response to a very slow, very steady buildup of stress in the crust, not to any sort of sudden underground burp. They are unpredictable, or at least very difficult to predict, because faults teeter on the edge of failure for years, decades, maybe much longer. We don't understand what happens to finally push a fault over the edge.

Away from subduction zones and active hot spots, the earth's strata are less dynamic. No giant slabs of seafloor sink into the mantle; no long-lived upwellings of magma push their way through the crust. Below about ten kilometers (six miles), the pressure in the crust is high enough to render rocks mostly impermeable to fluid flow. Nothing moves easily at these depths. But things do move. Fluids do exist.

The inner workings of the earth's crust remain mysterious because almost all of our information is based on indirect inference. It is possible to drill into the crust and peer directly at the strata; possible but not easy, and definitely not cheap.

The deepest drill hole to date reached a depth of twelve kilometers (over seven miles) in the Baltic Shield. The Kola Superdeep Drillhole was a massive scientific undertaking by the former Soviet Union. Drilling began in 1970 and included a number of boreholes branching away from a central hole. In 1989 researchers hoped to drill to a depth of 13.5 km, but had to stop at 12.2 km when temperatures in the hole reached 180°C (356°F)—significantly hotter than they had expected, and at the ragged edge of what the drill bit could withstand. Still, it was an impressive accomplishment. The experiment yielded a few surprises, including the fact that the earth's crust was more heterogeneous than expected, also considerably more wet. Based on previous work, researchers expected the rocks below about five kilometers to be different from those at more shallow depths. In particular, they expected to reach a volcanic rock known as basalt. The drill rig did reveal a change at the expected depth, but not what researchers had expected. The deeper layers were metamorphic rocks that had started out similar in composition to the more shallow layers, but had been profoundly altered by intense heat and pressure. The process of metamorphosis had shed water, not unlike the processes that generate fluids in subduction zones. Although buoyant, the water stayed trapped deep in the Baltic Shield, unable to make its way up through impermeable overlying strata.

Again, in a place like the Baltic Shield, the earth's deep strata change much more slowly than in an active subduction zone. But minerals do undergo transformations, and the mantle beneath the crust continues to convect, and fluids deep in the crust will sometimes be disturbed.

That earthquakes can disturb fluids in the crust is beyond dispute. A mountain of evidence as well as common sense tells us that when huge blocks of the crust shift, the underground plumbing system can be affected. Well levels can rise or fall; springs can dry up, or be created; geysers can change their tunes. The world's most famously regular geyser, Old Faithful, erupted reliably every sixty-four minutes on average until a large earthquake struck near Yellowstone in 1959. Now it erupts on average every ninety minutes. Two things control the behavior of any geyser: the geometry of subsurface conduits and the supply of subsurface fluids. A large earthquake can easily alter either or both of these.

But could the equation go the other way? Could fluids in the crust have something to do with earthquake nucleation? Laboratory studies say the answer might be yes. As rocks are subjected to stress they develop extensive small "microcracks" as they reach a breaking point. As these cracks grow and coalesce, water and gases might find themselves able to move. When trapped fluids are subject to increasing stress the pressure of the fluid, the pore pressure, increases—a process that can cause rocks to fracture. Once cracks are generated, gases could be released, making their way up into groundwater, changing its chemical signature. Significant volumes of water might be liberated from formerly confined aquifers, potentially changing the electrical conductivity of the rock. In the words of earthquake hydrology expert Michael Manga, "scenarios of this kind have led to the not unreasonable expectation that hydrological, hydrogeochemical, and related geophysical precursors might appear before the occurrence of large earthquakes."

Manga cites the Soviet Vp/Vs studies as examples of such (possible) precursors. Other intriguing documented precursors include the rise in the local water table three days before a magnitude 6.1 earthquake in central California, and changes in groundwater chemistry in the days before the 1995 magnitude 7.2 Kobe, Japan earthquake.

In the late 1990s a team of researchers led by Dapeng Zhao used earthquake waves to probe the earth's crust in the vicinity of the Kobe earthquake. Looking at waves in the Kobe region, Zhao's team found unusual wave velocities across a three-hundred-square-kilometer region around the initiation point, or hypocenter, of the Kobe earthquake. Similar anomalies have now been observed in the locations of other large earthquakes, suggesting that earthquakes nucleate in regions where fluids exist deep in the crust.

Other lines of research, on the edge if not the fringe of the mainstream community, focus on the possibility that earthquakes are preceded by the release of subterranean gas, in particular radon. The idea of so-called radon anomalies dates back to the 1970s. An inert gas, radon is present in small quantities in rock formations; this gas, so the theory goes, is released as rocks undergo increasing pressure in advance of a large earthquake. A number of studies have purported to show increased radon release prior to large earthquakes. All such studies played the usual game, identifying precursors in noisy data after earthquakes have occurred—the familiar bugaboo of prediction research.

In 1979 scientists at Caltech pointed with optimism at apparent radon anomalies prior to magnitude 4.6 and 4.8 earthquakes in southern California on January 1 and June 29 of that year. But the larger magnitude 5.7 Coyote Lake earthquake struck northern California near garlic-capital Gilroy on August 6 of that same year, unheralded by significant fluctuations on any of the monitoring instruments, including radon recorders, in operation near the fault. In early October seismologists at Caltech announced that new anomalies had been detected in two southern California wells, one in Pasadena and one about 15 miles east, in the town of Glendora, the previous June and July. The observation garnered attention in scientific as well as media circles, with headlines striking an ominous chord. At 4:45 p.m. local time on October 15 southern California was rocked by a serious quake, but the magnitude-6.4 Imperial Valley event struck in the desert near the U.S.-Mexico border too far from the reported radon signals for anyone to suggest a connection. Radon levels, it began to appear, had a knack for bouncing up and down for reasons other than impending earthquakes.

In a few cases, for example a study done by Egill Hauksson and John Goddard looking at earthquakes in Iceland, the statistical significance of radon anomalies was considered. These researchers started with a data set that included twenty-three earthquakes and fifty-seven total observations of radon emissions prior to these quakes (i.e., observations were made at multiple sites at any one time.) Looking at the fifty-seven observations they found nine apparent radon precursors prior to earthquakes, but forty-eight cases where precursors were not detected prior to earthquakes. They also found a high rate of "false alarms"—apparent radon precursors that were not followed by earthquakes. They concluded that there was no statistically significant evidence for a relationship between radon anomalies and earthquakes.

Following the 2009 L'Aquila earthquake in central Italy an Italian researcher pointed to his earlier prediction, which had been ignored. Close inspection revealed a familiar tale. Although in this case the prediction had apparently been borne out, it was based on observations—small earthquakes and anomalous radon release—that we know do not provide the basis for reliable prediction.

In the minds of most seismologists, none of the early radon studies are compelling when analyzed rigorously. In other scientific circles research continues apace. In recent years Russian physicist Sergey Pulinets has been a leading proponent of such research. Considering the consequences of radon release, Pulinets argues that the gas would generate aerosol-size particles in the air. These aerosols, as Helmut Tributsch argued, could account for other sorts of (purported) observed precursors, including so-called earthquake clouds, decreased conductivity of the air, and temperature (thermal) anomalies. Pulinets is a serious guy; his theories are generally sound and cannot be dismissed as junk science. The problem (or at least, one problem) lies with the observations the theories purport to explain. Pulinets and his cohorts show images of apparent thermal anomalies, or earthquake clouds, prior to past large earthquakes. But here again, one is predicting earthquakes after the fact. Pulinets's skeptics ask, where are the statistics that prove these anomalies to be anything more than the usual fluctuation of noisy data? Where are the rigorous statistics that establish a link between the purported anomalies and earthquakes?

Here it is important to return to an earlier point, namely that it is possible that some of the signals are real—that is, that fluids and gases do have something to do with earthquake nucleation, and do generate the signals that have been identified as earthquake precursors, but are of no practical use for prediction because the precursors aren't reliable indicators of impending quakes. Leading seismologist Hiroo Kanamori, who has reviewed prediction literature and debates at some length, suspects this might be the case. Considering examples like the Matsu-shiro swarm he considers it possible if not likely that at least some identified earthquake precursors, both hydrological and electromagnet, are real in the sense of being caused by a process that is associated with the subsequent earthquake activity, but that the processes are so varied and complicated one might only ever be able to identify precursors after the fact—not ever to develop a reliable prediction method based on these precursors.

Thus we are left with some suggestion that fluids or gases in the crust might have something to do with the occurrence of large earth-quakes. But if, how, and when these fluids move or otherwise conspire to cause rocks to fracture, we have ideas but except in a small handful of cases no real understanding. For all of the intriguing hydrological and hydrogeochemical precursors that have been described in the sci-entific literature, Manga concludes that definitive and consistent evi-dence remains elusive.

When we look back at the earthquake activity in the Bohai Sea re-gion prior to the Haicheng earthquakes, our intuition tells us that something was brewing deep in the crust before the earthquake hap-pened. Our intuition might be right, it might be wrong. Fluids in the crust might have something to do with earthquake nucleation, but we are a very long way from understanding the observations, let alone the theories. The percolation processes that played out in the earthquake prediction arena after Haicheng are considerably more clear.

CHAPTER 8

The Heyday

What is 10 inches high in some places, covers more
than 4,500 square miles, and worries the hell out
of laymen and professionals alike?
—GEORGE ALEXANDER, *Popular Science*, November 1976

*T*he full story behind the Haicheng prediction remained murky for
several decades, and the story did not grab immediate headlines
around the world. The scientific community was, of course, well aware
that an earthquake of portentous magnitude had struck northern
China. And even before Haicheng struck, western seismologists were
aware of the active Chinese prediction program. In the months following the earthquake, western scientific circles began to buzz. On
February 27, 1975, just weeks after the earthquake, then chairman of
the earth and planetary sciences department of MIT Frank Press wrote
an op-ed article that appeared in a number of papers around the country, telling readers that "recent results in the Soviet Union, China,
and the United States indicate that measurable changes that occur in
the earth's crust signal earthquakes. Further development can lead to
predictions of great earthquakes many years in advance of their
occurrence."

On April 27, 1975, Clarence Allen was quoted in the *Los Angeles
Times* making balanced but fairly favorable comments about the Chinese earthquake predication program. By August, a lead editorial in
the same newspaper began, "Seismologists are getting close to a remarkable achievement: the ability to forecast with increasingly greater
accuracy the location, time, and magnitude of earthquakes."

Seismologist Robert Hamilton, then chief of the Office of Earthquake Studies in the Interior Department, set out to collect published reports as well as information from Robin Adams's visit. In November of 1975 he stepped forward with a public announcement that Chinese authorities had predicted the Haicheng earthquake. Hamilton's version of the story again parroted the party line: a prediction made at top levels, relayed to local governments, a general evacuation of Haicheng and Yingkow on the afternoon of February 4. Thus the script, written in the East, was established in the West.

Hamilton's announcement garnered modest but not enormous media interest in late 1975. In early November 1975, Hamilton gathered U.S. Geological Survey scientists and government officials from nine western states. The order of business: to discuss how scientists and officials would respond if a credible prediction were to be made in the United States. Clearly if one gathers top officials to discuss how they would respond to an earthquake prediction, the implicit message is clear: not only scientists but also public officials were looking to the day, sooner rather than later, when earthquakes would be predicted routinely. Indeed, Hamilton stated as much—that accurate predictions might be possible in as little as a year. "In California," Hamilton said, "where prediction may come first, most residences are of wood-frame construction, which stands up well in earthquakes. . . . Certain hazardous buildings should be evacuated, but these could be specifically identified."

The media continued to take note. On November 9, 1975, readers of the *Fresno Bee* were informed that "as early as one year from now, Gov. Edmund G. Brown Jr. could get a startling telephone call. On the other end of the line would be a scientist from the United States Geological Survey. His message: Within 30 days the San Francisco Bay area or Los Angeles or any of the other cities along the San Andreas fault can expect a major earthquake. That's how close American geologists are to earthquake prediction."

Hamilton was, of course, well aware of the general sense of excitement that remained in the air in the aftermath of the 1971 Sylmar quake. He knew that many top experts in the field felt a genuine sense of optimism that prediction was right around the corner. He knew

that the USGS had dispatched one of its top young scientists to Garm
to better understand apparently promising developments in the Soviet
Union.

But he knew more than that.

In the aftermath of the Sylmar quake, USGS scientists Jim Savage
and Bob Castle had teamed up to "see if there might be something in
the geodetic record." In particular, they wondered if the earth's surface
had undergone any unusual warping either before or after the earthquake. In the 1970s the business of geodesy still relied on ground-
based surveying measurements, sparse and imprecise data compared to
today's GPS and InSAR data. But then as now, a wealth of surveying
data was available for southern California. With early surveying data
subtle horizontal warping was difficult to resolve, but such data are
pretty good for telling us if the ground has moved up or down. Savage
and Castle figured that if there were interesting precursory signals to
be found, southern California was the place to look for them. They
collected available surveying data, and set about figuring out what the
earth's surface had been doing both before and after 1971.

Castle and Savage found apparent elevation changes in the San Fernando Valley area in the years prior to 1971. Their paper was published
in the prestigious journal *Geology* in November of 1974. They also
found a significant uplift over a swath of the Mojave Desert north of
the Los Angeles metropolitan region, more-or-less centered along the
San Andreas Fault. They carefully considered the then-known sources
of error that might have crept into their analysis, and concluded that
the uplift signal was real.

Neither Savage nor Castle went pedal-to-the-metal in touting the
Mojave signal as a harbinger of a future great quake. Castle is the first
to acknowledge that his colleague Jim Savage is "as conservative as
they come" in his interpretation of observations. Earth scientists have
been known to occasionally take observations at face value rather than
turning them inside out with a critical eye to make sure they are real.
Jim Savage has built a career—a reputation—on his critical eye. In
1998 he was awarded the Medal of the Seismological Society of America, the society's highest honor. Working with Bob Castle his conser-

vatism stood him well, but even this most critical of scientists was not entirely immune to the optimism of the day. The final sentence of the 1974 paper reads, "If the uplift [in the Mojave] is, in fact, ultimately determined to be a premonitory phenomenon associated with continuing strain on the San Andreas Fault, it suggests an impending earthquake of at least the magnitude of the San Fernando event." In other words, the signal might not be a precursor, but it could be; and if it were, then the impending quake would likely be big. Although Castle and Savage remained careful to avoid rash or otherwise unjustified statements, optimism clearly was in the air. The Palmdale uplift quickly generated a buzz within the community.

If there is enough electricity in the air it doesn't take much of a spark to cause a conflagration. Other scientists were quick to identify the Mojave signal as a likely precursor, and to develop theories to explain it. Among them was Max Wyss, who in 1977 published a paper in *Nature* arguing that the uplift could be explained by the process of dilatancy, the process whereby the volume of a material actually increases (dilates) when it is subjected to shear stress. This was the same process that had been first described several years earlier to explain the Soviet V_p/V_s results. Wyss put together a scientifically credible interpretation of Castle's observation, concluding that the crust around the San Andreas Fault had begun to swell in 1961, the signal growing to include the Sylmar region by 1971, and continuing to grow in subsequent years. Thus Wyss explained the 1971 quake as a "sideshow of a future repetition of the 1857 San Andreas break, which had a magnitude > 8."

Wyss's paper wasn't published until April of 1977; the political conflagration ignited much sooner.

Was Bob Hamilton aware of Castle's results when he stepped forward with bold public statements in November of 1975? "You bet he was," Castle replied.

Among the worker bees in the earth science community Castle's observation generated excitement and optimism tempered by conservatism and skepticism. Among ranks of top scientific leadership—those individuals with the right combination of scientific acumen and political leadership to push programs forward—one found less ambivalence. Frank Press, by that time the head of the U.S. Geological Survey's Earthquake Studies Advisory Panel, first learned of the bulge at a December, 1975 meeting of the panel. In January 1976, Frank Press and several officials from the USGS were invited to the White House to give a presentation, an opportunity they used to make their pitch for increased funding for prediction research. Press took the opportunity to mention the Palmdale bulge to Vice President Nelson Rockefeller, who expressed interest and asked for more details. Still, through 1976 the Ford administration balked at the funding increase that Press and others were pushing for.

The scientific paper describing the bulge was published in *Science* in April 1976. But days after a magnitude 7.5 earthquake struck Guatemala on February 4, 1976, killing more than twenty thousand people and grabbing headlines around the world, the USGS released word of the bulge to the press. On March 11, 1976, USGS scientists took their concerns about the uplift to the California State Seismic Safety Commission. Top scientists presented a balanced view, with renowned geologist Robert Wallace noting that "[they were not making] a prediction, at least not now; the uplift is simply an anomaly we really don't understand." Caltech's Clarence Allen was also careful to avoid an alarmist tone: "We can't say that an earthquake is going to occur tomorrow, next year, or ten years from now." He also noted, reasonably enough, "I think there's reason to be concerned. There's always been reason to be concerned." Neither Castle nor Savage was invited to the meeting.

Tempered words notwithstanding, the message came through loud and clear: the scientific community was concerned. Regardless of what exactly was said at the meeting, the fact that they had brought their results to the state commission—and, effectively, to the public— spoke volumes. The predictable headlines followed: "'Bulge' on Quake Fault 'May Be a Message,'" readers of the *Long Beach Independent Press-Telegram* were told.

In April of 1976 Frank Press spoke at a special session of the American Geophysical Union, describing the successful Haicheng prediction, which he described as having been based largely on an observed uplift of the earth's crust. In a talk titled "A Tale of Two Cities," Press went on to discuss the "worrisome uplift" that had been detected along the San Andreas Fault. He further "cited reports that seismic researchers in southern California claim their work is being slowed by insufficient funds," noting that the region of concern is not well instrumented with sensors that could detect movement of the ground surface, unusual gas concentrations, changes in well level, and such. Press chose not only his words but also his implications carefully.

By the spring of 1976, reports from California had begun grabbing national headlines. In the hands of veteran science writer George Alexander, Castle's "uplift" had become the "Palmdale bulge." "What is 10 inches high in some places, covers more than 4,500 square miles, and worries the hell out of laymen and professionals alike?" Alexander asked in the opening line of a 1976 *Popular Science* article.

Some top scientists continued to voice optimism tempered by caution. Cast in the prominent skeptic role was Charles Richter. Richter, who developed his famous scale at the beginning of his career in the 1930s, saw the Palmdale bulge play out toward the end of his career. An enthusiastic and effective life-long spokesman, Richter did not make a point of expressing his skepticism in a public arena. But in scientific circles his views were well known, and by the accounts of those who knew him, drove a wedge between him and scientists who had been his closest colleagues.

Still, many scientists were understandably intrigued by reports from China, and viewed the Palmdale bulge with a measure of real concern. Other tantalizing, seemingly credible results began to surface as well. In 1976 Caltech researcher Jim Whitcomb stepped forward with a prediction that a moderate earthquake would strike north of Los Angeles before April of 1977. This prediction was based not on the bulge, but rather on results that seemed to indicate that the rocks in the vicinity of the San Andreas Fault had undergone a significant change over the preceding years—the same type of change that researchers in the Soviet Union claimed to have identified. Whitcomb had made an earlier prediction using the same method. In December 1973 he predicted that a magnitude 5.5 or greater quake would strike east of Riverside, California, within three months. Although he was later credited by some with a hit after a magnitude 4.1 quake struck on January 30, 1974, scientists today would not regard this as a successful prediction. By 1976 earthquake prediction had become a hot topic, and Whitcomb and his prediction made it all the way to the pages of *People* magazine, in another article written by George Alexander. "Earthquake prediction, long treated as the seismological family's weird uncle," he wrote, "has in the last few years become everyone's favorite nephew."

As the Palmdale bulge story rippled across the media and generated heightened anxiety across southern California, Hamilton, Press, and others worked in political circles to translate public concern into political support.

The efforts to launch a more comprehensive earthquake program were by that point nearly a decade old. It had been back in 1965 that a panel headed by Press had recommended to the White House that a $137 million program was urgently needed to predict earthquakes and mitigate earthquake risk. Looking back from 1975 one can trace the decade-long struggle to get such a program off the ground. The USGS had positioned itself in a lead role with its creation of an earthquake center in Menlo Park, in effect creating the program without the budget. In 1969 USGS director William Pecora headed an interagency committee that wrote a proposal for a ten-year national earthquake hazards program. This report, viewed by many as overly narrow and, more notably, not inclusive of key players in the hazard mitigation community, was not warmly embraced. In particular Pecora's plan focused exclusively on earthquake science, leaving hazard assessment and earthquake engineering out in the cold.

A year later, Karl Steinbrugge pulled together a more comprehensive proposal that was embraced as a consensus-building document. Steinbrugge was a respected researcher who, like Frank Press, had become influential in policy circles. During the 1970s the men were both tapped as advisors to the Office of Management and Budget. Steinbrugge's 1970 report avoided the shortcoming of the Pecora report. Whereas Pecora had honed in on detailed budgets, leaving all participants immediately aware of the pie to be split and their potential piece of it, Steinbrugge avoided the subject of money completely. It was a plan everybody could love.

Steinbrugge's report is recognized as having provided a key infusion of momentum, but nothing happened overnight. The National Academy of Sciences appointed a panel of experts to weigh in on the question of earthquake prediction. Their charge: to determine whether prediction was in fact a credible goal for earthquake science. By the early 1970s, panel chairman Clarence Allen indicated in a preliminary report that the panel would answer the question in the affirmative.

In the absence of a bold new federal program the USGS had continued its efforts to launch its fledgling earthquake program. The process by which an agency like the USGS requests money involves a protracted dance between the agency, Congress, and the Office of Management and Budget. In effect, an agency asks for what they are allowed to ask for. The U.S. Geological Survey moreover has a range of different programs, and has to weigh proposed requests for new money for one program against the needs of other programs. In 1975 the USGS requested an additional $16 million to launch a comprehensive earthquake prediction and preparedness program. In response the Office of Management and Budget authorized only $2.6 million for earthquake programs but did not increase the overall USGS budget. The new money would have to be squeezed out of existing USGS and National Science Foundation (NSF) budgets. Thus the nascent earthquake center was left to keep itself afloat by scrounging for loose change under the cushions.

Then the public and public officials found themselves facing what the experts portrayed as a real and present danger. Robert Olson, executive director of the California Seismic Safety Commission, made a push for support from the federal Disaster Assistance Administration to study earthquake hazards in southern California. The proposal for a new earthquake program had also found champions, and fresh momentum, on a national stage.

In fall of 1976 the NSF and USGS produced what became known

as the Newmark–Stever report, the namesake of Guy Stever, President
Gerald Ford's science advisor, and Nathan Newmark, a renowned engineer from the University of Illinois. In the words of Bob Wallace, the
panel decided it was "time to get busy with actual programs to do
something." Twelve years after the great 1964 Alaska quake, and eleven
years after a national program was first proposed, the finish line was
finally in sight.

In early 1976 Senator Alan Cranston and Representative Philip
Burton, both from California, had introduced legislation for an Earthquake Disaster Mitigation act that would allocate $50 million per year
for at ten-year national earthquake program. It was not Cranston's first
attempt to introduce such legislation; he had previously teamed up
with Charles Mosher from Ohio and Representative George Brown
from California to build an advocacy coalition to push legislation forward. Bills had been introduced in 1972, 1973, and again in 1974, yet
the program remained stalled on the launch pad.

By 1976 earthquakes were on the national radar screen. Memories
of the 1971 Sylmar quake had not yet faded. The New Madrid Seismic
Zone in the central United States had sprung to life with a moderate
shock on June 13, 1975, a quake that was not large enough to cause
damage but was widely felt throughout the region. The disastrous February 4, 1976 Guatemala earthquake had struck uncomfortably close
to home for those in the western United States. In June of that year
National Geographic published a pair of articles, one on the earthquake
disaster, the next describing the optimism that earthquake prediction
might be within reach. Before the year was out the USGS published a
glossy report on this earthquake, noting in the introduction that "there
is much to learn from this earthquake that is directly relevant to the
problem of reducing earthquake hazards in the United States." Meanwhile an ominous bulge imperiled southern California and, while the
Chinese had already reaped the rewards of their bold investment in
earthquake prediction, the U.S. scientific community lacked the funds
to try to assess the situation properly.

As the Cranston bill negotiated its way through Congress the Ford
administration convened a panel headed by Nathan Newmark to for-

mulate a working plan for the new program. Populated by a mix of scientists and engineers, the panel considered the extent to which the new program should focus on prediction, as opposed to efforts to mitigate earthquake risk. Geologists Joe Ziony from the USGS and Lloyd Cluff from the consulting firm Woodward-Clyde found themselves in the minority, arguing in favor of emphasizing research to improve our understanding of earthquake rates and shaking, and to strengthen the resilience of buildings and infrastructure. They did not carry the day. As Cluff recalls, fellow panel member Clarence Allen did voice reservations about touting earthquake prediction as a feasible, short-term goal. Allen was optimistic that progress could be made, but also foresaw an uphill battle on an extraordinarily difficult scientific problem. Others on the panel made a full-court press for prediction as a centerpiece of the new program.

In the end the Newmark panel recommended that funding for earthquake hazard studies be increased from a current $20 million to $75–$105 million by 1980. They recommended a balanced program, with support for hazard studies as well as engineering research, but specifically recommended significant increases for earthquake prediction research. Newspaper articles through the summer and fall of 1976 described the program as a hazard reduction effort, yet emphasized the "stepped-up funding of research into earthquake prediction." When the bill failed a vote in the House of Representatives the *Northwest Arkansas Times* ran a big article on September 29. The headline read, "Earthquake Hazards Act Fails in House." The first sentence of the article informed readers that "earthquake prediction failed, by a tie vote of 192-192, to achieve the two-thirds majority needed to suspend the rules and pass the Earthquake Hazards Reduction Act." The grammar was dubious but the point was clear.

Cranston's legislation fell a whisker shy of success in 1976, but the tide turned with the presidential election in November. In January of 1977, Jimmy Carter brought a nuclear engineer's sensibilities with him to the highest office in the land. By February of 1977 Frank Press was able to speak from the pulpit not only as one of the leading seismologists of the day but also as "the man in line to become President

Carter's science advisor." Back at Caltech, where he had been on the faculty earlier in his career, as part of a distinguished lecture series, he spoke of the need for research and international cooperation to develop effective earthquake prediction.

By May of 1977 the Earthquake Hazard Reduction Act (EHRA, Public Law 95-124) had been passed unanimously by the U.S. Senate. By October it had passed both houses of Congress. The 1977 congressional act defined the goals of a national earthquake hazard reduction program, and tasked the Office of Science and Technology Policy (OSTP) with the responsibility of developing an implementation plan. The OSTP was headed at the time by none other than Frank Press, who by that time had indeed stepped into the role of science advisor to Jimmy Carter. Press tapped Steinbrugge to head the effort to pull the plan together; Steinbrugge in turn tapped a team of experts representing a broad range of disciplines. Thus did the years of planning—of strategizing, of report writing, of politicking—finally pay off. Thus did, in the words of historian Carl-Henry Geschwind, "earthquake hazard mitigation [become] entrenched within the federal government apparatus."

Looking back one can now reflect with a measure of perspective on the long journey that culminated in the launch of the National Earthquake Hazards Reduction Program (NEHRP). In a retrospective 1997 article, Christopher Scholz noted that prediction research was in fact never a big part of the program. But reading newspaper articles from the mid-1970s, it is clear that, in the eyes of the public as well as public officials, the promise of earthquake prediction was clearly the driving impetus. At least some perceived the rhetoric as hype. A 1976 editorial cartoon in the *Los Angeles Times* depicted a chicken all-atwitter, its head thrown away from its body, squawking, "The earth is quaking! The earth is quaking!" But the success of the journey had everything to do with the idea that an infusion of federal dollars would give the public what they wanted, namely reliable earthquake prediction.

It is interesting if not fruitful to reflect on motivations. When scientists set out to launch new programs, the line between genuine conviction and opportunism can start to blur. From the outside looking in it is hard if not impossible to tell the two apart. But if conviction was in some cases bolstered by expedience, there is no question the conviction, and the excitement, was real. The understanding of earthquakes had, or so it seemed, taken a quantum leap forward in recent years.

When Peter Molnar set off for Garm to learn more about apparent Soviet success with prediction he brought with him a scientist's innate skepticism, but also curiosity and a measure of optimism. He was, after all, among a generation of young scientists who had just watched a bold new idea—the theory of plate tectonics—win out over established conventional wisdom on the part of the old guard. It was not a generation lacking in confidence. Molnar's optimism might have evaporated during his visit to Garm, but his report to Frank Press, at that time on the faculty at MIT, failed to squelch the latter man's conviction that the Soviets had achieved real success with their earthquake prediction program. Other young scientists who spoke with Press in the mid-1970s were less sure about their senior colleague's conviction. For his part, earthquake prediction skeptic Lloyd Cluff saw the debate as very much shaped by political tensions of the day. The United States was behind our two Cold War nemeses, China and the Soviet Union, in research of critical importance to life safety. Clearly it was incumbent on the United States to catch up. When Cranston's bill was voted down by the House, a front page article in the *Los Angeles Times* quoted the manager of the bill, describing it as an effort to close "the earthquake gap between us and the Chinese."

The extent to which prediction was touted as purely a matter of political expediency, we will never know. The extent to which the "genuine optimism" was fueled by wishful thinking if not worse, we will never know. One thing we do know, and can consider: what became of all of the optimism once the national program was launched.

CHAPTER 9

The Hangover

Short-term earthquake prediction represents a more
difficult scientific problem than most of us thought 5 yr
ago when the National Earthquake Hazard
Reduction Program commenced, and our progress
has not been as rapid as initially hoped.

—CLARENCE ALLEN, 1982

*A*lthough sold largely on promises of prediction, Congress's stated
purpose for the National Earthquake Hazard Reduction Program
(NEHRP) was, "to reduce the risks of life and property from future
earthquakes in the United States through the establishment and maintenance of an effective earthquake hazards reduction program." From
the beginning Congress recognized that more than one federal agency
could contribute to the NEHRP mission. Today there are four primary NEHRP agencies: the Federal Emergency Management Agency
(FEMA), the National Institute of Standards and Technology (NIST),
the National Science Foundation (NSF), and U.S. Geological Survey.
All four agencies play important roles, but through the 1960s and 1970s
the USGS maneuvered wisely. Established in 1879 with a primary mission of mapping mineral resources, the USGS has since the beginning
taken the lead role within NEHRP with earthquake research and hazard assessment.

The USGS Earthquake Hazards Program, supported by NEHRP,
includes internal USGS programs as well as a substantial program of
grants to the researchers in universities. The USGS Earthquake Hazards Program today touts many successes: improved national seismic
hazard maps, greatly improved earthquake monitoring, development

of sophisticated earthquake notification systems, a better understanding of earthquakes and the shaking they produce, and so forth. When people today click on a Web page and find the magnitude, location, and shaking distribution of an earthquake just minutes after it happens, they are seeing the fruits of NEHRP. The USGS earthquake program moreover bore tremendous scientific fruits, with numerous contributions that have advanced our understanding of earthquakes. But perusing the Web sites that today tout the program's numerous accomplishments, one facet of earthquake science is conspicuously absent: earthquake prediction. Looking back one can see how sentiments regarding earthquake prediction evolved following the launch of NEHRP.

The buzz over earthquake prediction did not fade away immediately in scientific or public circles. A spattering of articles continued to appear, in California newspapers in particular, touting seemingly promising results. In 1980 Caltech scientists spoke to the media about the possibility that an upsurge in radon concentration might presage future earthquakes on the San Andreas or San Jacinto faults. Articles in 1980 and 1981 described an apparently ominous confluence of strange signals from the earth, not only radon but also other changes, including a rise in the water table in parts of San Bernardino and the activity of springs near the San Andreas Fault. John Filson, then director of the USGS office of earthquake studies, told the *Los Angeles Times*, "These anomalies do give us some concern," although he added, "We don't see anything right now to warrant a change in our assessment of southern California's seismic status."

Aftershocks of continuing excitement notwithstanding, in retrospect the decline in optimism was precipitous. The sense of hope generated by the prediction of the Haicheng earthquake took a hit even before NEHRP got off the ground. At a minimum, the deadly M7.8 Tangshan earthquake, which struck in July of 1976, provided a reality check for the scientific community and the world. If the Chinese had successfully predicted one earthquake, clearly they had not unlocked the secret of reliable earthquake prediction.

By the end of the 1970s the famed Palmdale bulge had also developed serious cracks. A number of prominent scientists began to discuss and analyze serious measurement errors lurking in the apparently ominous Palmdale bulge signal. Castle and a number of his colleagues at the USGS remained steadfast in their insistence that sources of errors had been considered carefully, and could not explain away the uplift. Other scientists argued to the contrary, eventually generating a debate that became as intense, as heated, as personal, and as nasty as any earth science debate in recent memory. Ross Stein, one of the first USGS scientists to break ranks, twice found paper bags full of dog excrement in his USGS mailbox. David Jackson, one of the first scientists to carefully explore the sources of error, adds, "Some people told me I was full of crap, but nobody sent me any."

By 1981 veteran science writer Dick Kerr wrote an article in *Science*: "Palmdale Bulge Doubts Now Taken Seriously." A few years later the bulge had all but foundered. Ironically, after the dust settled and all sources of error were considered carefully, the bulge never did disappear completely; eventually it was explained as a consequence of the 1952 Kern County earthquake. But as a harbinger of doom it had clearly not lived up to expectations. As it had since 1857, the San Andreas Fault remained locked.

Neither had Jim Whitcomb's predicted earthquake occurred. By the late 1970s the once promising Vp/Vs method was foundering in general. Back in 1973, both the *New York Times* and *Time* magazine had been quick to credit Yash Aggarwal and his colleagues with a successful prediction based on the method. Upon sober reflection the small earthquake looks easy to explain as part of a sequence, including earthquakes as large as magnitude 3.6, that had begun in mid-July. The closer seismologists looked at the method, the more they realized that apparent changes in Vp/Vs resulted from the fact that measurements at different times were made using different sets of earthquakes. When one looked for changes using quarry blasts, which unlike earthquakes were controlled and repeatable sources, the anomalies disappeared. From the beginning some skeptics within the seismology community had harbored suspicions about these studies. If earthquake prediction skeptics had felt pressured in the 1970s to bite their tongues in the face of optimism from their colleagues, by the end of the decade tongues were loosened by increasing evidence that earlier studies were flawed.

IS DILATANCY DEAD?

Chris Scholz, the scientist who first proposed the dilatancy theory to explain the Vp/Vs observations, points out that the theory itself remains sound. Laboratory experiments tell us that when rocks are subjected to increasing stress, cracks do form, and rock volume does increase. These changes will give rise to changes in Vp/Vs. In Scholz's view, the early observational studies did not disprove dilatancy; rather, the results lacked sufficient precision to prove *or* disprove the theory. He further points out that a number of studies designed to look for evidence of dilatancy were fundamentally flawed because they focused on types of faults, including the San Andreas, for which signs of dilatancy will be extremely difficult to detect. Whether the observational basis for dilatancy will rise from the dead as seismologists develop tools to peer into the earth with increasingly surgical precision remains to be seen. In Scholz's view, at least, the demise of dilatancy has been greatly exaggerated.

Another infamous chapter in the annals of earthquake prediction research played out just as the Palmdale bulge was deflating. Two scientists, Brian Brady and William Spence, came forward with a specific prediction that a major earthquake would strike Peru near Lima in 1981. Brady, who had earned a PhD in geophysics from the Colorado School of Mines, hailed from the United States Bureau of Mines (USBM), where his research focused on rock bursts in mines. Founded in 1910, the USBM was for years at the forefront of research in the minerals and mining fields. By the time Brady splashed onto the scene the agency's glory days were waning as the USGS fortunes were waxing. Spence worked with the USGS office in Golden, at the time a more operational arm of the USGS than the high-powered research center in Menlo Park.

Although Spence was an energetic supporter of the prediction to which his name was attached, his colleague Brian Brady was from the beginning the driving force behind the prediction. Brady's interest in earthquake prediction began with research to understand rock bursts: the spontaneous, often violent fracture of rock that sometimes occurs when deep mine shafts are drilled, reducing pressure on neighboring rock. Brady concluded that rock bursts are preceded by characteristic, identifiable patterns of very tiny earthquakes—essentially cracking in the rock prior to the explosion, and had some apparent success predicting rock bursts. He began to develop a theory to explain how a so-called inclusion zone develops, evolves, and ultimately collapses. The fundamentals of the theory, which Brady laid out in a series of articles in the mid-1970s, were essentially conceptual, which is to say, described by ideas, not by equations. Nonetheless Brady identified three classes of precursors that were expected to precede failure of a system: long-term indicators of impending failure, short-term indicators, and very-short-term indicators.

Appealing to a principle known as scale invariance, Brady reasoned that there should be no difference between the processes that control rock bursts and the processes that control earthquakes. Seismologists generally believe in scale invariance, insofar as small and big earthquakes appear to differ only in size, not in the nature of underlying physics that controls earthquake rupture. Whether the physical processes that control rock bursts are the same as those that control earthquakes is another question entirely. In Brady's mind, conventional ideas about earthquake nucleation were off base because they didn't "address the fundamental problem of *how* the fault gets there in the first place." The seismologist, in turn, regards this as problematic: the earth's crust is riddled with faults that have all existed for a long time. Fault zones might continue to develop and "grow," but almost without exception, earthquakes occur on preexisting faults. Earthquakes moreover represent an entirely different type of failure from rock bursts. As scientific principles go, scale-invariance is a good one in many cases, but it does not predict any commonality between rock bursts and earthquakes if they are controlled by different processes. It does not allow one to, as Brady did in one paper, grab an equation that describes the energy change associated with a collapsing void space and use that equation to draw conclusions about precursors to earthquakes.

BREAKING ROCK

Because rock bursts occur when mining reduces the confining pressure on neighboring rock, Brady's lab experiments, which involved subjecting unconfined samples of granite to increasing stress, were regarded by seismologists and rock mechanics experts as not useful for elucidating the process of failure during an earthquake. Applying the lessons learned from such experiments to earthquakes one ran headlong into a fundamental disconnect: faults deep in the earth might be relatively weak compared to the surrounding crust, but, unlike mine walls, they remain under enormous confining pressure.

In retrospect one is struck by the disconnects as well as the conceptual nature of Brady's papers. But during the prediction heyday of the 1970s more than a few prediction papers were published that do not impress the modern reader as entirely rigorous. Also, today as well as in the 1970s, long and difficult papers do sometimes slip through the peer-review process in spite of serious shortcomings. When this happens the ultimate judgment is often made by the community: the paper is ignored.

Brady nonetheless pressed forward. He had started to look at earthquakes in the early 1970s. Looking at foreshock activity prior to the 1971 Sylmar, California, earthquake, he believed could identify the same patterns as those that preceded rock bursts. He concluded that the earthquake could have been predicted, had the patterns been identified beforehand. On October 3, 1974, a magnitude 8.1 earthquake struck southwest of Lima, Peru, killing seventy-eight people and causing heavy damage. Brady turned his attention to the region and began to be concerned that he was seeing a pattern of activity that pointed to a much larger earthquake in the near future. In particular he was concerned that the aftershock sequence died down abruptly following a magnitude 7.1 aftershock on November 9. Based on Brady's observations and theories, this pattern suggested that the preparation phase for a great earthquake had begun.

Over the following several years Brady continued to analyze the data and refine his theory. By August of 1977 he formulated his initial specific prediction, namely that a magnitude 8.4 earthquake would strike near Lima in late 1980. After further work the prediction became even more alarming. In an internal memo he wrote in June of 1978 that "the forthcoming event will be in late October to November, 1981 and that the magnitude of the mainshock will be in the range 9.2 +/− 0.2. This earthquake will be comparable to the 22 May 1960 Chile earthquake." Eventually the magnitude of the predicted earthquake climbed further, to 9.9, which, in the words of Clarence Allen, "certainly didn't add to the prediction's credibility." In July of 1980 a top USGS official wrote an internal memo to John Filson, the head of the USGS Office of Earthquake Studies. "FYI and FYA(musement)," he wrote, "I talked to Krumpe at AID/OFDA who seriously informed me Brady's latest prediction re: Peru earthquake is . . . foreshocks begin 23 Sept. 1980; mainshock 316 days later. I just love precise estimates, don't you?"

For his part, Spence's scientific contribution was largely in assessing past earthquake activity and the structure of the offshore subduction zone, which he concluded could indeed host a much larger event. However, he continued to champion Brady's work and the prediction. In 1978 he had coauthored a meeting report on a conference on earthquake prediction; Brady's theories are featured prominently in the report.

By 1978 Brady was in communication with a leading scientist in Peru as well as top officials at the U.S. Geological Survey. Notwithstanding the keen interest in earthquake prediction on the part of the USGS, their officials and top scientists were underwhelmed by Brady's prediction. At a key meeting in the spring of 1979 scientists at the USGS Menlo Park office pointed to the lack of published papers explaining the theory in full, let alone papers demonstrating its validity.

The meeting also included representatives of the USBM, the USGS Golden office, scientists from the lead geophysical institute in Peru, and the U.S. Office of Foreign Disaster Assistance (OFDA). The OFDA, whose mission is to promote hazard mitigation and disaster preparedness worldwide, was in favor of *not* dismissing the prediction.

As the discussion, which at times revealed hints of interagency turf skirmishes, continued in the United States, word of the prediction started to reach the Peruvian public in late 1979. In February of 1980 the president of the Peruvian Red Cross visited the director of OFDA to appeal for U.S. aid, including a list of key preparedness items. Word of the appeal leaked to the Peruvian media, the attentions of which focused immediately on one particular item on the list: a request for one hundred thousand body bags. The news splashed with tsunami-force intensity across Peruvian airwaves and newspapers.

By the beginning of 1981 talk of the prediction permeated all layers of Peruvian society. Many pointed to the prediction as the cause of a significant drop in foreign tourists, then as now a lifeblood of the economy. The Peruvian Civil Defense was overwhelmed by requests for information.

In January of 1981 the Peruvian government asked the National Earthquake Prediction Evaluation Council (NEPEC) to evaluate the Brady-Spence prediction. Back in 1976, as the push for NEHRP had finally neared the finish line, the U.S. Congress had formally authorized the National Earthquake Prediction Evaluation Council, a panel of experts who would be available to "review predictions and resolve scientific debate prior to public controversy or misrepresentation." The council, chaired by Clarence Allen, met in Golden, Colorado, on January 26 and 27, 1981, to review the Brady prediction. The event, attended by TV crews and other members of the media, was an exercise in deep frustration for all concerned. Brady had counted on spending two days explaining his prediction; Allen suggested that he and Spence wrap up their presentation in a total of five hours.

Brady proceeded to try to explain his unorthodox, highly conceptual theories to the panel. By Spence's later account, Brady "got his back up" after he tried to explain his theory and found himself cut off at every turn. The transcript of the hearing reveals that midway through the first day Brady launched into a long discussion of his most recent theories, which attempted to incorporate magnetic forces, thermodynamic stability, and equations that Brady likened to Einstein's field equation. The theories were again highly conceptual, what impressed the committee as a headache-inducing jumble of ideas. As Brady progressed to talk of cosmology, Clarence Allen cut him off. "There has been a request from the committee that we stop this because no one is understanding what is going on." The committee included some of the top minds in earthquake science, with expertise ranging from seismology to rock physics. The group of nine very smart people found themselves at a loss. Several committee members noted that they had received no prior information about the very complex theories Brady was talking about. Allen observed, "I hope you appreciate that you are turning members of the Council off to some degree partly by coming up with information that we have not been advised about despite the fact that it has been known for months that this meeting was going to take place, and the prediction [was made] over a year ago."

Council member Jim Savage told Brady, "This isn't a criticism, but I say, I don't think members of the panel have understood what you are saying. You are wasting your time, and you had better get to something that is perhaps within our comprehension, or present it more thoroughly so we can understand it." Fellow council member David Hill, who had always marveled at Savage's ability to see to the core of complex theoretical arguments, immediately understood what Savage was really saying—that Brady wasn't making any sense.

By the end of the first day all of the NEPEC members had, in fact, concluded that Brady's theories were, in effect, off the rails—not simply flawed, but neither rigorously formulated nor well-grounded in theory or observation, and certainly not ready for prime time. They saw flaws in the theories based on scale invariance, including an error that had been pointed out by leading seismologist Keiiti Aki in a paper that was published in 1981, but had been circulated in draft form earlier. When Brady started to expound on a connection between prediction and Einstein's theory of relativity, committee members saw the discussion heading off into outer space. Their comments through the first day—"gosh, we're sorry, we don't understand . . ."—followed scientists' time-honored tradition of maintaining politeness in the face of science they do not consider credible.

ON UNORTHODOXY

Breakthroughs in science do sometimes happen when researchers find unexpected connections between apparently disparate scientific fields. Although Brady never published a paper explaining in full the theories he presented to the NEPEC panel in 1981, in 1994 he published a conference proceedings article arguing that rock fracture could be understood in the context of so-called critical point theory. A critical point essentially is reached when a system finds itself in limbo between two states—for a fault, between a state where stress is building and one in which stress is being relieved. It is within the realm of possibility that insights into earthquake processes will some day be derived from this, or one of the other highly unorthodox approaches that Brady explored. Such insights would, however, have to be derived from rigorous development of theory. Conceptual science can lead to interesting ideas, but it can take you down a path to perdition, and at best can only take you so far.

At the end of the very long day panel member Lynn Sykes decided to fly home. Another panel member, Rob Wesson, came to the conclusion that it would be necessary to challenge Brady more directly. His and his colleagues' resolve was strengthened, and their commitment to politeness greatly weakened, by evening and morning news stories that in their view portrayed the NEPEC committee as having been unable to comprehend the novel theories presented by a brilliant, maverick young scientist. An article published in the Tuesday morning *Rocky Mountain News* described how Brady had used "elaborate mathematical formulas" to develop theories that "the times, locations, and magnitudes of certain types of earthquakes can be predicted with extreme accuracy." The article further noted that "panel members admitted they couldn't understand his mathematics and asked him whether the theories were essential to his prediction." The article went on to quote Paul Krumpe, science advisor to USAID/OFDA, as saying, "This is science in the making. It might as well be Einstein up there." The same tone is evident in the concluding paragraph of one article published after the second day of the hearing, "Brady contended his application of Einstein's theory of relativity to breaking rock and earthquakes was an essential aspect of his prediction and complained repeatedly throughout the hearing that panel members didn't want to hear about it."

The second day of the meeting was thus marked by a different tone. The committee could not understand the fundamentals of the theory or the logical connection between the theory, the data, and the specific prediction. They pressed Brady on specifics at every turn. As Brady tried to describe the nucleation zone in terms of an A_H and an A_C "cosmological horizon," one panel member pressed for Brady to write down the equations for these "horizons" in the simplest possible case. When Brady equivocated in reply, Wesson asked, not entirely politely, how the panel could be sure that Brady wasn't getting his prediction from the Tibetan Book of the Dead.*

James Rice, a leading rock physics expert who was not part of NEPEC but had been invited to attend the meeting as a nonvoting consultant, pressed Brady on his understanding of rock fracture research. "Fracture has been studied," he observed, "and it has been a scientific subject for a large number of years. There are many problems . . . which are reasonably understood, and solved, and I am trying to make some contact with that literature and the body of knowledge and the concepts that you are putting forth here."

---

*The transcript of the NEPEC hearing is riddled with mistakes. Scientific terminology is mangled: "defamation" instead of "deformation," "Ace of H" instead of "A_H" ("A sub H"), etc. The transcript also misattributes the "Book of the Dead" line to Bob Engdahl, to the consternation of Rob Wesson, the committee member responsible for the line.

The committee pressed Brady on other details of the prediction as well, asking repeatedly to see equations that were still not forthcoming. Brady eventually did produce a very simple equation showing that the time scale for precursory earthquake patterns was proportional to the size of the impending quake. The committee pounced on an apparent contradiction. The precursory pattern that Brady had identified prior to the 1971 Sylmar earthquake had played out over about eight years, yet, while it had been just seven years since the purported pattern had developed in Peru, Brady was predicting a much larger earthquake. Brady replied that "[it] has a lot of things going into it, and you have temperature coming in and you have a precursor loading. . . ." Rob Wesson replied in turn, "I guess that is why we need the equation."

As the hearing drew to a close committee member Barry Raleigh remarked on Brady's failure to present a convincing theoretical explanation to explain his prediction. He also noted that Brady's "previous public work has got errors in it, which we have not discussed, but it doesn't give me great confidence . . . in the so-called work you are presenting here today." He concluded that in his opinion, "the seismicity patterns that you purport to show here are clearly ad hoc, and I see no relationship to theory."

Following an executive session the council noted that it was impossible to say that a major earthquake would not happen on any given day, but their evaluation of the specific prediction was unequivocal. They had "been shown nothing in the observed . . . data, or in the theory insofar as presented, that lends substance to the predicted times, locations, and magnitudes of the earthquakes." The NEPEC pronouncement was transmitted through the U.S. State Department to the president of Peru: the prediction did not merit serious consideration.

Back in Peru, however, concern only continued to mount. Concern was to some extent stoked by Paul Krumpe. In April of 1980 Rob Wesson, writing a colleague in France, noted that "the supporters of the prediction (Brady; Bill Spence, USGS, Golden; and Paul Krumpe, USAID) seem to share an almost messianic belief and fervor." Krumpe continued to champion Brady in bureaucratic circles, explaining that "Brady's current hypothesis appears unique in that it departs from accepted Einstein physics (Field Theory) and classical rock mechanics. He offers a comprehensive rational physical explanation for the following elements which, regardless of scale, contribute to rock failure, rock bursts, and the occurrence of earthquakes." In contrast to the leading experts in seismology and rock physics who had weighed in during the NEPEC meeting, Krumpe alone, it seemed, had an appreciation for Brady's revolutionary theories. In July of 1981, after the incident defused, Clarence Allen, a scientist well known among colleagues for speaking in measured words, wrote to the then-head of USAID to express concerns that Krumpe, "had taken it upon himself not only to embrace the Brady prediction, but actually to aid and abet Dr. Brady in its promulgation." Allen went on to say that "Mr. Krumpe seems to have perceived his proper role as protecting the brilliant, young martyr from the big, bad scientific establishment." Others wondered about the extent to which Krumpe was seeking to further his own ambitions. NEPEC committee members' views of Krumpe's professional judgment were not improved by his later association with the Montana doomsday cult established in 1990 by Elizabeth Clare Prophet. Brady himself viewed Krumpe's personal beliefs as extremist.

The details of Brady's prediction, including the timing and magnitudes of major shocks, had been a moving target since the beginning; by April of 1981 he predicted a major (magnitude 8.2 to 8.4) foreshock on June 28 and a later, magnitude 9+ mainshock.

By June 1981 William Spence formally disavowed the prediction after a predicted foreshock failed to occur, saying in a memo that he always felt the term "Brady-Spence prediction" overstated his role. Brady remained steadfast, although he did say he would withdraw his prediction if the first predicted shock failed to occur.

In April of 1981 the U.S. Embassy in Lima had enjoined John Filson, then deputy chief of the Office of Earthquake Studies at the USGS Reston, Virginia headquarters, to visit Lima in June. The cable noted, "Embassy staff strongly believes that visit to Peru by Dr. Filson at this time would go long way to help allay public fear and put Brady's predictions in proper perspective."

Filson heeded the call, arriving in Lima on June 25. Having been well steeped in the debate for several years the visit was a revelation for him. "I had no idea," he wrote in a report, "of the level of anxiety and concern these predictions had caused in Lima. During my stay, every newspaper contained at least one front page story about Brady; property values have fallen drastically; many who could afford it left town for the weekend, and the people at the hotel where I stayed said their bookings were down to about one-third normal." Filson recalls the eerie quiet that he met, walking around what he knew should be a vibrant urban community. At the American ambassador's house where he had been invited one evening, the ambassador's wife served tuna fish sandwiches for dinner. The household staff, including the cook, had gone home to be with—perhaps die with—their families.

June 28 passed quietly, at least in geological terms. Filson's four-day visit, during which he emphasized the formal NEPEC rejection of the prediction, was front-page news. The Peruvian newspaper *Expreso* ran a full page, front-page headline that day: "NO PASO NADA" ("Nothing Happens"). The subtitle quoted the reaffirmation by noted seismologist John Philson [*sic*] that the prediction had not been scientifically credible. So too were the statements of Peruvian authorities, which had formerly been somewhat ambivalent but now expressed unequivocal rejection of the prediction.

In a July 9 report on his trip, Filson expressed concern that the date of the predicted earthquake had shifted at least three times since May, most recently to July 10. "If he is allowed to continue to play this game . . . he will eventually get a hit and his theories will be considered valid by many," he wrote. Filson's fears were soon put to rest. Brady reportedly began a draft of a formal retraction on July 9, although it was not sent out until July 20. The prediction and the episode were put to bed. In the annals of earthquake prediction, the Brady–Spence prediction is not regarded as a shining moment.

Thus, with one highly visible prediction fiasco and the promise of earlier studies clearly not borne out, did the heady days of the early 1970s give way to an earthquake prediction hangover. For John Filson the episode marked a definite turning point. He had been tapped to serve as head of the USGS Office of Earthquake Studies, initially in an acting capacity, in February of 1980. At the time the program was less than three years old, and had little plan or definition. In its internal and external programs the USGS had supported a wide range of research, in Filson's words, "everybody all over the place." In the midst of this free-for-all era Filson's phone occasionally rang in the wee hours of the morning, a researcher calling with breathless excitement to report anomalous readings from their instruments. He would later check back with researchers, only to be told that, oh, the signal hadn't been real after all, a capacitor had blown.... Geophysicist Mark Zoback, then at the USGS, watched the hangover play out from a vantage point closer to the scientific trenches. He saw optimism fade in the face of "an endless progression of negative results." The more closely scientists scrutinized apparent precursors, the less convincing the results became.

But neither the USGS nor the U.S. seismology community gave up on prediction research entirely. In particular, and driven by Filson's newfound conviction that a more focused approach was in order, in 1984 USGS scientists formulated the prediction that would go down in infamy, namely that a moderate quake would occur on the San Andreas Fault near Parkfield within four years of 1988. The USGS moved forward with a focused experiment after the state of California agreed to put up matching funds for instrumentation. The National Earthquake Prediction Evaluation Council met and endorsed the science on which this prediction was made. The USGS invested substantial time and resources installing instrumentation to catch the earthquake red-handed. Looking back, some younger scientists point out that the Parkfield experiment was launched in the hopes of catching earthquake precursors, but on a forecast rather than a prediction; that is, on the idea that quakes recur regularly on certain faults, not on any specific indication that an earthquake was imminent. It is perhaps a fine distinction. Enthusiasm for earthquake prediction programs did not disappear overnight, perhaps especially in the organization that had become the lead agency for earthquake research under NEHRP.

However, enthusiasm definitely faded. In 1982 Clarence Allen wrote a short article for the *Bulletin of the Seismological Society of America.* The first sentence of the abstract: "Short-term earthquake prediction represents a more difficult scientific problem than most of us thought 5 yr ago when the National Earthquake Hazard Reduction Program commenced, and our progress has not been as rapid as initially hoped." Although Allen went on to proclaim himself agnostic on the question of whether earthquake prediction would ever be possible, and to conclude that prediction remained a worthy and important research avenue, he tallied a number of early results that had failed to hold up to closer scrutiny and discussed new lines of research pointing toward earthquakes as unpredictable phenomena.

The tone of articles in the popular media began to follow the lead of scientists like Allen. A July 11, 1982 article in the *San Francisco Chronicle-Telegram* described progress made on "the earthquake problem"; the progress cast in terms of preparedness, not prediction. The article went on to discuss scientists' earlier work on prediction, including investigation of anomalies in radon level and animal behavior, noting that "they are still probing these anomalies, but their connections to earthquakes have been more fickle—or at least harder to find—than expected."

In 1976 the *Los Angeles Times* published over 150 articles about earthquakes, including several dozen on prediction and the move to launch a new federal program. By 1985 the paper ran a total of 126 articles on earthquakes, eighty-five of which appeared in the month following the devastating Mexican earthquake on September 19 of that year.

The U.S. earth science community still wasn't sure what to make of the supposed Haicheng prediction, but the catastrophic and unpredicted Tangshan earthquake ended every Haicheng sentence with an exclamation point. Maybe Haicheng, if not exactly a fraud, had been a fluke. In the early 1970s earthquake prediction had been right around the corner. In earthquake hazard circles, at least in the United States, a different mantra emerged: resources should be focused not on prediction efforts that might never pan out, but on research that stood to help us better understand earthquake shaking and earthquake hazard. Hazard assessment, not earthquake prediction. Hazard assessment might not be sexy, and it might not have been what the public wanted to see, but NEHRP had been launched and professionals in the earthquake hazard mitigation business set about creating a program to mitigate risk.

Today one can learn about the history of the NEHRP program on a Web page at www.nehrp.gov. The site notes that "changes have occurred in program details in some of the reauthorizations, the four basic NEHRP goals remain unchanged." These goals are listed: (1) Development of practices for earthquake loss reduction, (2) improve techniques for reducing earthquake vulnerabilities, (3) improve hazard identification and risk assessment methods, and (4) improve the understanding of earthquakes and their effects. Although earthquake prediction research arguably falls under the last of these, the word *prediction* does not appear on the Web page. Without question, the program has been directly responsible for real strides toward all of these goals. Progress toward prediction, however, slowed quickly and eventually came to a virtual standstill.

As the 1980s wore on the National Earthquake Prediction Evaluation Council, which had been pressed into service to weigh in on a number of predictions, including the Brady-Spence prediction and the Parkfield prediction in 1984, found itself with less and less to do. In 1995 NEPEC quietly disappeared. In the words of USGS geophysicist Michael Blanpied, the committee simply found itself "out of work."

That the USGS reestablished NEPEC in 2005 perhaps serves to illustrate that, while skepticism still runs deep, the hangover has at least abated to the point that scientists can once again talk about the "p-word" in polite company. But the 1970s are not ancient history. Today's young and midcareer scientists did not experience these times firsthand, but they have certainly heard the stories. Apart from prediction, research seismologists have seen just how slippery statistics can be when we seek to characterize earthquake patterns. We are collectively older; and one likes to think, at least, that we are collectively wiser.

CHAPTER 10

Highly Charged Debates

Many strange phenomena precede large earthquakes.
Some of them have been reported for centuries, even
millennia. The list is long and diverse: bulging of the
Earth's surface, changing well water levels, ground-
hugging fog, low-frequency electromagnetic emission,
earthquake lights from ridges and mountain tops,
magnetic field anomalies up to 0.5% of the Earth's
dipole field, temperature anomalies by several degrees
over wide areas . . . changes in the plasma density
of the ionosphere, and strange animal behavior.

—Friedemann Freund,
Journal of Scientific Exploration, 2003

*W*hile optimism for the prospect of reliable earthquake prediction
was fading in the U.S. other dramas were playing out on other
shores. On February 24, 1981 a magnitude 6.7 earthquake struck Ath-
ens, killing sixteen people and injuring thousands. Just as the Sylmar
quake had mobilized American scientists a decade earlier, this disaster
galvanized the scientific community in Greece. Soon after the earthquake a pair of solid-state physicists, Panayotis Varotsos and Kesser
Alexopoulos, began monitoring electrical signals in the earth, reasoning that electrical currents would be generated as stress reached a critical point prior to a large earthquake. A third colleague, Kostas Nomicos, soon joined the team, and the so-called VAN earthquake
prediction method was born. And the debate soon began.

The VAN team has continued to develop and refine their method,
in recent years including analysis of earthquake patterns as well as

electrical signals. At the heart of the method, however, is the claim that "seismic electric signals," or SES, are generated in the earth prior to earthquakes. These signals, or so the theory goes, are not detectable everywhere in the vicinity of an impending earthquake, but only at "sensitive sites."

The details of the VAN method are the stuff of which lengthy and dense journal articles are made. And not only that: entire books have now been written on the method. To avoid stumbling into a statistical thicket of tangled trees, one can focus on two aspects of the forest.

First there is the fundamental nature of modern science. Science moves forward—science can only move forward—on the basis of testable hypotheses. A scientist develops an idea for how the world works based on observed phenomena. The idea appears to do a good job of explaining the observations. The idea is however useless unless it leads to predictions about future observations, predictions that can be tested to either prove or refute the hypothesis. With the concept of "sensitive sites" the VAN method edges not just down a slippery slope but also over a cliff. That is, no amount of negative evidence—the absence of SES prior to large earthquakes—can ever disprove the hypothesis, because any and all negative results can be dismissed as having been recorded at insensitive sites.

Then there is the fact that the VAN method has been around for over two decades, long enough to have established a track record. Critics of the method sometimes point to the absence of a solid theoretical framework. In fact one wouldn't necessarily need to understand how an earthquake prediction method works if one could demonstrate beyond a shadow of a doubt that it does work. But does VAN work? Proponents are quick to point to successes: apparently successful predictions of a handful of moderate earthquakes.

Closer scrutiny reveals a different story. The signals interpreted as SES are not uncommon; passing trains can generate very similar (if not identical) signals. Identification of significant SES is thus to some extent an art as much as a science. Meanwhile, Greece is known for a propensity to swarmy seismic activity. Which is to say, while we know that earthquakes in any area tend to cluster, Greece is among the areas where this tendency is especially strong. After a moderate earthquake in Greece, it's a reasonably good bet that additional quakes will follow. Looking carefully at the VAN track record in the early 1990s, Francesco Mulargia and Paolo Gasperini found that VAN predictions tended to follow rather than precede significant earthquakes. Other studies have reached similar conclusions, namely that the limited apparent successes touted by VAN proponents can be attributed to their ability to capitalize on earthquake clustering. Questions have further been raised, by seismologist Max Wyss among others, about outright falsifications on the part of VAN proponents.

But here again, when some seismologists, including Hiroo Kanamori, consider the overall geological setting of Greece, they consider it not implausible that fluids in the crust have something to do with earthquake activity in the region. Greece is in an extensional environment, which is to say, the crust is being pulled apart by plate tectonic forces: the Gulf of Corinth is widening at a rate of 1–1.5 centimeters (1/2–1/4 inch) per year. In extensional environments, magma as well as other fluids in the crust will likely find pathways along which to migrate upward in the crust, and the movement of fluid within rocks can generate electrical signals. Thus, while Kanamori appreciates the cracks, so to speak, in the arguments put forth by Varotsos and his colleagues, he again wonders if the SES signals might indeed be linked to processes at work in the earth. To his mind the highly charged debate over prediction is an unfortunate distraction from what should be the business of science, namely investigation of these processes.

Earthquake prediction efforts in Japan have also followed a different trajectory from those in the United States. As in the United States, the Japanese government's interest in prediction dates back to early optimism during the 1960s. Whereas this optimism had largely dissipated in the United States by the late 1970s, in 1978 the Japanese government enacted the Tokai quake prediction law, abbreviated DaiShinHo in Japan. Echoing the earlier Haicheng saga, concern for a future great (magnitude 8) Tokai earthquake was stoked by a series of small quakes in the region in the 1970s. That a future great earthquake in this region, about one hundred miles from Tokyo, is inevitable is beyond

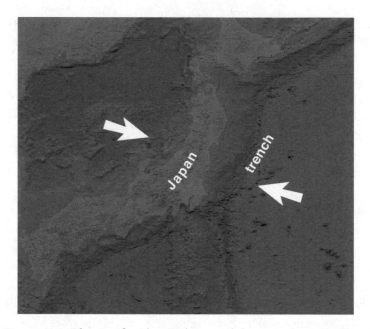

Figure 10.1. The Pacific Plate sinks, or subducts, at the Japan Trench. (Image courtesy of USGS.)

dispute. The Japanese islands are what geologists call an island arc, geological flotsam built up along an active subduction zone (fig. 10.1). The last devastating quake along this zone struck in 1923, the magnitude 8.3 Kanto earthquake, which claimed over one hundred thousand lives as a consequence of building collapse and the firestorms that followed the quake.

The last great quake struck the Tokai region back in 1854; it had an estimated magnitude of 8.4. The long historical record in Japan reveals the usual story of irregular clocks. The average time between great quakes is 110 to 120 years. Again the earth exhibits significant variability around the law of averages, but the Tokai segment is Japan's southern San Andreas Fault—the one that might not go tomorrow, but that surely has to go sometime.

When the Bohai Sea region in China started to rumble in the late 1960s, Chinese seismologists had no prior basis for expecting a big quake in that particular region. In general, we know that northern China experiences fewer damaging earthquakes than the southern reaches of the country, where the crust is being squeezed and stressed by the ongoing push of the Indian subcontinent northward. When a number of quakes occurred in the highly active Tokai region in the 1970s, concern naturally followed. By 1978 the government passed a law mandating not only preparations for, but also prediction of, the next great Tokai earthquake.

DaiShinHo was, from the start, much more than just an earthquake prediction program. It channels significant resources to earthquake preparedness efforts, for example construction of tsunami walls in harbors. Only a small fraction of the program's resources are earmarked for research, including prediction efforts. Still, the pot of money has continued to exist, and, earthquake science funding being what it is, seismologists in any country can rarely afford to ignore any funding source that might continue to support their research. The Japanese earthquake science community grew by leaps and bounds, in size as well as sophistication, in the 1970s and 1980s. Here too, early optimism for prediction faded among reputable seismologists as apparently promising early results were not borne out by later work.

But unlike the United States, Japan has had sixty years of nearly unbroken one-party rule, and tremendous bureaucratic inertia. Thus DaiShinHo lives on. In the view of Caltech-educated seismologist Robert Geller, vocal earthquake prediction skeptic who has spent his career doing research and making waves on the faculty of Tokyo University, reputable academic seismologists in Japan have for years walked a fine line, writing proposals to continue sensible earthquake investigations, doing their best to avoid outright lies as they describe how their work advances a cause they don't believe in. "At this moment," Japanese physics professor Masayuki Kikuchi said in 1994, "the majority of seismologists in Japan believe earthquake forecasting is either impossible or very difficult."

On January 17, 1995, one year to the day after the 1994 Northridge, California, earthquake, a magnitude 6.9 quake struck the Hanshin (Kobe) region of south-central Japan. Whereas Northridge caused massive dollar losses but claimed only a handful of lives, over five thousand people were killed in the Hanshin quake, mostly in the city of Kobe. Buildings as well as infrastructure took a heavy hit. For a country that had prided itself on earthquake preparedness, Hanshin was a psychological as well as geological shock. Well beyond the shores of Japan, the world took note. It was the kind of damage and death toll one expected to see in the developing world, not an economic power-house like Japan—and certainly not in a country that had arguably led the world in its earthquake preparedness efforts.

Support for earthquake science and preparedness skyrocketed. Much of this money went in sensible directions. Some of it went to-ward research on the VAN method, not to seismologists but rather earth scientists from other disciplines who had by that time developed an interest in the method. These efforts were never warmly received by the Japanese seismological community, but from the outside look-ing in, one could easily get the impression that earthquake prediction research was alive and well in Japan.

In general, since the hangover days of the 1980s, earthquake predic-tion research has remained out of favor in the United States as well as, for the most part, the United Kingdom and most of Europe. Through-out this time, however, it has been pursued with more optimism and enthusiasm by scientists, and sometimes government officials who don't have an appreciation of the views of their top scientists, in other countries.

IRAN

An estimated 35,000 people lost their lives when a large earth-quake struck northwest Iran on June 20, 1990. Following this di-sastrous event, the earthquake professional community in Iran pushed successfully for a new risk reduction program. The result, the long-term, "Iran's Strategy of Earthquake Risk Reduction," led to a major program to evaluate and strengthen the earthquake safety of the country's school buildings. Although improved resilience re-mains a challenge in a country where earthquake hazard is high and historical construction is culturally valuable but seismically vulnerable, thousands of school buildings have been retrofitted or replaced—and thousands more will be retrofitted or replaced in future years—under the program. Whereas damaging earthquakes in other countries have clearly led to a clamoring for earthquake prediction, Iran has been notably effective in channeling available resources to risk reduction. Asked how Iran managed to steer a different course, leading Iranian earthquake engineer Mohsen Ghafory-Ashtiany replies that, while there is interest among scientists in earthquake prediction, he and his colleagues take great pains to avoid talking to policy makers or the public about prediction, lest they be inspired to chase an elusive goal at the expense of one that engineers know will make real contributions to societal resilience.

In recent decades scientists who study earthquakes have parted company along disciplinary as well as international lines. As earthquake prediction fell out of favor among scientists who call themselves seismologists it has been embraced more enthusiastically by (some) scientists with expertise in different fields, for example rock magnetism and solid-state physics.

In particular, researchers have continued to explore the age-old idea that earthquakes are preceded by detectable electromagnetic precursors, and to develop methods such as VAN that are based on electromagnetic observations. Electromagnetic signals come in various shapes and sizes. The VAN method is based on electrical signals; other types of anomalies have been proposed, and purportedly observed. In the days prior to the magnitude 6.9 Loma Prieta earthquake south of the San Francisco Bay Area, a magnetometer in operation close to the epicenter recorded what appeared to be a highly unusual signal. The instrument had been installed by Stanford University professor Anthony Fraser-Smith to monitor variations in the earth's ultra-low-frequency, or ULF, magnetic field. Fraser-Smith was looking to study solar magnetic storms, the usual cause of ULF variations. On October 5, 1989, his instrument near the town of Corralitos recorded an increase in ULF intensity. On October 17, three hours before the quake, the signal increased even more dramatically, reaching an amplitude about thirty times larger than normal. No similar excursion had been recorded during the two years of operation prior to this time. The results were touted in *Science* less than two months after the earthquake, and published in 1990.

Among the true believers in earthquake prediction circles, few cows are as sacred as the Corralitos record. Corralitos itself is a tiny (population 2,430), relatively affluent enclave in a remote part of the Santa Cruz Mountains. It is not on the road to anywhere. The average Californian probably has no idea that Corralitos exists. The average earthquake prediction aficionado, amateur or professional, knows all about Corralitos. This one short snippet of data launched a thousand ships. The idea of electromagnetic precursors was scarcely new in 1989. But when the Corralitos record appeared in the pages of *Science*, here, it seemed, was a truly serendipitous, truly compelling observation that the key to earthquake prediction lay in monitoring not earthquakes, but the earth's magnetic field.

Sensibly, Fraser–Smith and others set out to install more instruments in earthquake-prone regions. Nothing like the Corralitos signal has been seen since 1989. In 2006, geophysicist Jeremy Thomas got intrigued by the famous record and began comparing it to the handful of records from similar instruments at other locations. He realized that in many ways, the details of the Corralitos record were similar to the details of records from thousands of miles away, consistent with expectations since ULF variations are generally due to solar rather than local changes. He then found a simple way to make the different records match even more closely, essentially by assuming that the time stamp and gain of the instrument had gone awry on October 5. In short, he showed that dramatic anomaly could be explained by an instrument or data processing malfunction. The idea is scarcely outlandish. In the 1980s geophysical monitoring equipment was hampered by limitations in computer processing and memory. Raw data from the Corralitos instrument weren't even stored, just smaller snippets of processed data. This particular instrument was moreover in a remote location, routinely tended by a volunteer, not frequently visited by a scientist or technician. The logbook that describes the maintenance of the instrument has never been made available.

Thomas's analysis does not prove that the anomalous signal was due to an instrument malfunction. Sacred cows die hard. At the 2007 International Union of Geodesy and Geophysics (IUGG) meeting in Italy scientists from countries around the world presented talk after talk purporting to show correlations between various electromagnetic signals and earthquakes. Thomas's paper, presented in the same session, was not received warmly. Nor did USGS geophysicist Malcolm Johnston do much for international goodwill among scientists when he presented an overall critique of prediction research based on electromagnetic precursors.

The details of the research presented at the 2007 IUGG meeting are mind numbing. The VAN method, which alone has been the subject of much exhaustive debate in the literature, is only one among many types of electromagnetic-based prediction methods now under consideration by scientists in any number of countries. A detailed discussion of theories, results, and debates would be impenetrable to any reader for whom earthquake prediction research is not a life calling.

Among the torchbearers in this community, solid-state physicist Friedemann Freund stands tall. His research, including laboratory investigations, provide compelling evidence that battery currents can be created when certain rocks are subjected to stress—what has come to be known as Freund physics.

The path to Freund physics began with chemistry. In his investigations of the chemical and physical processes inside crystals, Freund found that small amounts of water get incorporated in a crystal lattice when a mineral crystallizes under the pressures found deep in the earth's crust. The water does not remain in the lattice as H_2O, but forms pairs of two different molecules that each contains one oxygen atom, one hydrogen atom, and one silicon atom. When these pairs form, the hydrogen atoms each takes one electron from each oxygen atom, creating positively charged oxygen atoms. Such atoms—that is, charged by virtue of chemical reactions—are known as ions. These oxygen ions then bond with silicon atoms, which renders them essentially stuck.

Although Freund "did not think much of this discovery" for years, he eventually began to wonder if he might have the key to purported observations of electromagnetic signals prior to earthquakes. Freund realized that under normal conditions rocks are very good insulators, which is to say, very bad conductors of electrical currents. But when rocks are subject to stress, as they are deep in the earth, the bonds between the oxygen ions and silicon can be broken, generating charge carriers that are known as positive holes, or p-holes ("foles") for short.

As the theory goes, p-holes exist commonly in rocks deep in the earth, for the most part minding their own business. If the rock is then subjected to additional stress, both p-holes and free electrons can start to flow. Up to this point, Freund physics is not conjecture. Although some solid-state physicists take exception with aspects of the theory, laboratory experiments done by Freund and others demonstrate that battery currents can be generated within some rocks, under certain conditions. The question is not whether charge carriers exist in rocks deep in the earth, but whether they have anything to do with earthquake processes.

Freund argues that the electrons can create currents that travel downward into the deeper, and hotter, parts of the lower crust, whereas the p-holes will generate different currents, which travel horizontally. Both types of currents will produce electric fields, and these will interact with each other in a way that produces electromagnetic radiation.

Further, as a result of currents activated by stress, some of the p-holes will be carried to the earth's surface, effectively leaving a large region of the surface positively charged. Such a disturbance could cause various sorts of mischief, including discharge phenomena that could account for so-called earthquake lights and even disturbances of the plasma in the ionosphere, ninety to 120 kilometers above the earth's surface. Recall that when Helmut Tributsch suggested in the late 1970s that charged aerosol particles, or ions, could account for a wide range of purported precursors, he struggled to develop a mechanism that could plausibly generate the currents that would be needed to generate ions. Freund physics provides a candidate mechanism. In the minds of some, observations of seemingly anomalous electromagnetic signals prior to earthquakes, as well as ionospheric disturbances, confirm that "Freund physics" is alive and well in the earth, and represents the long-sought key for earthquake prediction.

EARTHQUAKE LIGHTS

Around 373 BC a Greek historian wrote that, "among the many prodigies by which the destruction of the two cities Helice and Buris was foretold, especially notable were both the immense columns of fire and the Delos earthquake." Although the account does not make a chronology entirely clear, some point to this as the earliest known description of so-called earthquake lights. In any case many intriguing accounts have been written in recent centuries, phenomena ranging from flashes or bands of light, globular masses, fire tongues, and flames emerging from the ground. A small handful of blurry photos purport to show lights in a number of regions, including Japan and the Saguenay region of Quebec. A famous photo, showing bright luminosity splashed across a nighttime horizon, was snapped during the Matsushiro swarm in 1966. In our modern YouTube era, videos of earthquake lights have begun to surface, including several that show flashes of lights in the nighttime sky before the magnitude 7.9 Peru earthquake of August 17, 2007, and one of oddly shaped, rainbow-hued clouds in the sky minutes before the Sichuan earthquake of 2008 (figs. 10.2a and 10.2b; full-color images and videos can be found on YouTube.com). A number of theories have been advanced to explain earthquake lights, including piezoelectric discharge and, recently, Freund physics. Seismologists have struggled to make sense of this enigmatic, elusive phenomenon. We can't begin to describe it fully, let alone explain it, but neither can we dismiss it.

Figure 10.2a. Figure purporting to show earthquake lights during the 1966 Masushiro, Japan, swarm.

Figure 10.2b. A photograph, reportedly taken prior to the 2008 Sichuan earthquake, of unusual, rainbow-hued clouds.

Why don't seismologists believe it? It depends on whom you ask. In the minds of some Freundians, seismologists are close-minded, unwilling to accept the idea that scientists from disciplines other than seismology might have found the holy grail that has eluded us all these many years. At the 2007 IUGG meeting few card-carrying seismologists even showed up, a fact that did not go unnoticed—or unremarked on—by the scientists in attendance. Certainly the same accusation is made, loudly and often, by an active community of amateur predictors—individuals with varying degrees of training in the sciences and abiding convictions that, first, earthquakes can be predicted, and, second, the mainstream seismological community is determined to ignore them.

When seismologists consider Freund's arguments they tend to trip over three stumbling blocks. First there is the fundamental, intuitive but really rather tricky, assumption that stress increases, significantly and fairly suddenly, before a large earthquake happens. We simply have no direct evidence that the earthquake machine works this way, and quite a bit of evidence suggesting that it doesn't. To generate an earthquake, one of two basic things must happen: (1) stress must build, either gradually or suddenly, to the point that faults reach a breaking point; or (2) the fault must somehow weaken such that rupture is facilitated.

As Malcolm Johnston points out, electromagnetic signals themselves provide compelling evidence that stress does not build rapidly before a large earthquake. As a thought experiment, suppose that as stress builds in the earth, rocks do produce a series of physical changes that generate electromagnetic signals. The actual earthquake would be the culmination of the failure process. As such, surely the electromagnetic signals would reach a crescendo at the precise time the earthquake begins—"co-seismic" signals, in the technical parlance. (As an analogy, imagine a tree starting to crackle when stressed by high winds; the loudest crack will accompany the final break.) Earthquakes are known to generate electromagnetic disturbances; following a large earthquake in India in 1897 geologist Richard Dixon Oldham observed that earthquakes caused disruptions in telegraph communications, from which he concluded that earthquakes must generate currents in the earth. But modern instrumental recordings have yet to reveal any evidence for true co-seismic signals.

Magnetic instruments can and do record earthquakes. But upon close inspection, one finds, without exception, an absence of true coseismic electromagnetic signals. When an earthquake happens the grinding of rock against rock sends seismic waves into the crust. Earthquakes generate different types of waves, the fastest of which, the P wave, is essentially a sound wave traveling through the earth. P waves are pretty fast; they travel through the crust at about seven kilometers per second. At this speed a wave can travel around the world in about an hour. But P waves are pokey compared to electromagnetic signals, which travel at the speed of light—about 300,000 kilometers per hour. A signal generated anywhere on the planet would reach everywhere on the planet in less than one-tenth of a second. If, then, a burst of electromagnetic energy were generated when an earthquake begins, magnetic instruments around the world would detect a signal immediately, well in advance of the arrival of the first seismic wave. They don't.

As an aside, it would be a tremendous boon for us all if true co-seismic electromagnetic signals did exist. It would open the door, if not for earthquake prediction, then certainly for early warning. In contrast to earthquake prediction, the development of early warning systems rely on well-established science. In short, if one is able to tell that a large earthquake has started in a location, an advance warning can be transmitted to a remote location using modern telecommunications at the speed of light. The time difference between warning and shaking is essentially the time difference between lightning and thunder. The amount of warning depends on the distance from the earthquake—and, of course, the time that it takes for an early warning system to determine that a large earthquake is underway. If an earthquake is nearby, the thunderclap follows the lightning bolt almost immediately. But sometimes, early warning can be meaningful. In Japan, such a system has been in place for years, automatically bringing the bullet train to a halt if a large earthquake is detected. In Mexico City, where thousands lost their lives in 1985 due to an earthquake 350 kilometers away, alarm sirens now sound if a large earthquake is detected along the coast.

Effective early warning systems might be based on well-established physics, but in practice they are not easy to develop. During a magnitude 8 earthquake, the fault remains in motion for a couple of minutes. For effective early warning, one must record waves and know within seconds that the earthquake will be large. Then a warning must be generated and relayed in a way that will foster constructive response. The latter part of this equation is a concern.

If true co-seismic electromagnetic signals were real, the earth would essentially be its own early warning system. Rather than depending on sophisticated analysis of data from a network of seismographs standing on guard for large earthquakes, individual electromagnetic instruments in any location could detect unusually large signals and provide warning of impending shaking. A nice idea, if only co-seismic signals existed. They have never been observed.

Returning to the question of earthquake prediction, a second overarching criticism of earthquake prediction research based on electromagnetic precursors has to do with statistics. Recordings of electromagnetic signals are notoriously messy. Various types of instruments record various types of noise; as noted, VAN instruments record signals generated by trains, power lines, and so forth. They also record various types of natural signals from processes other than earthquakes. Typically, one has is left with a recording that varies all over the place, with spurious spikes as well as other excursions. Typically, one can find spikes and/or other strange signals that occur close in time to large earthquakes. The question, of course, is, does one have anything to do with the other?

Looking at purported observations of electromagnetic precursors of various stripes, the trained seismological eye looks for, and fails to see, the kind of rigorous statistical analysis that would prove these precursors are real, and significant. The trained seismological eye looks for, and fails to see, any evidence that anyone can identify a precursor before an earthquake happens, as opposed to looking back to identify the anomaly after the fact. With messy data and without rigorous statistical analysis a scientist does not have to be consciously dishonest to find apparent correlations where none exist.

A final criticism of Freund's theories stems from the fact that the earth's crust contains a lot of water, and water will cause currents to short-circuit—if they are generated in the first place. Other research-ers point out that water could actually serve as a conduit, bringing charged ions to the surface. For his part, Freund admits that certain questions remain to be answered, and says, "it is too early to say that earthquake prediction is just around the corner." But he adds, "I feel confident that the discovery of p-holes in rocks and their activation by stress represents a crucial step toward cracking the code of the earth's multifaceted pre-earthquake signals."

Indeed, while the idea of electromagnetic precursors is not entirely out to lunch these lines of research might never lead to reliable earth-quake prediction. Laboratory studies tell us that Freund physics is, at a minimum, not entirely black magic. Intriguing anecdotal accounts—and, recently, some photographic evidence—suggest that some earth-quakes are preceded by earthquake lights. Compelling theories and some evidence tells us that earthquakes might be preceded by hydro-logical disturbances, which could give rise to electromagnetic signals.

Precursory electromagnetic signals in the ionosphere have been re-ported in recent years by researchers who have analyzed data from the DEMETER microsatellite, launched in 2004 by the French space agency to investigate electromagnetic activity related to earthquakes. Many scientists remain skeptical of the results, and the authors them-selves emphasize that precursors can only be identified by stacking data from many earthquakes. Thus, even if the signals are real, there might be no hope of identifying an individual precursor in advance of an earthquake. Still, the studies have been rigorous enough to meet the bar for publication in respected peer-reviewed journals.

So why has it fallen to the community of scientists who consider themselves seismologists to be cast in the role of skeptic in so many earthquake prediction debates? Where the seismological community is coming from has a lot to do with not only what we have been through but also the lessons we continue to learn.

CHAPTER 11

Reading the Tea Leaves

> The chief difficulty Alice found at first was in managing
> her flamingo: she succeeded in getting its body tucked
> away, comfortably enough, under her arm, with its legs
> hanging down, but generally, just as she had got its neck
> nicely straightened out, and was going to give the
> hedgehog a blow with its head, it WOULD twist itself
> round and look up in her face, with such a puzzled
> expression that she could not help bursting out laughing:
> and when she had got its head down, and was going to
> begin again, it was very provoking to find that the
> hedgehog had unrolled itself, and was in the act of
> crawling away: besides all this, there was generally a ridge
> or furrow in the way wherever she wanted to send
> the hedgehog to, and, as the doubled-up soldiers were
> always getting up and walking off to other parts of the
> ground, Alice soon came to the conclusion that it
> was a very difficult game indeed.
>
> —*Alice in Wonderland*

Details of scientific debates aside, it is an interesting suggestion, that seismologists would turn a willful blind eye to truly promising earthquake prediction research—in particular to any researcher, or even any amateur, who started to establish a track record of clearly successful earthquake predictions.

So why are seismologists of all people persistently skeptical about, even dismissive of, what other scientists are convinced is promising earthquake prediction research? One simple answer is that, for all of the supposedly promising results, no such track record has been established, by anybody. Proponents of the VAN method claim a small number of supposed hits, but many more moderate and large quakes have struck with no warning. And, again setting all of the debate aside, in the years since 1989 scientists have yet to see another signal like the Corralitos record, let alone a successful prediction based on ULF precursors.

Another point is that the 1970s are not exactly ancient history. Many of today's leading senior scientists can recall vividly the hopes, the promises, and ultimately the disappointments of that era. More than a few senior scientists were directly involved with the research that helped fuel the sense of optimism that prediction was just around the corner.

The hangover of the 1980s left seismologists with a deep appreciation for the fact that seemingly compelling results will often vanish in a puff of smoke under closer scrutiny. In particular, seismologists have a hard-won appreciation of the pitfalls of looking back at data to look for anomalies after a large earthquake had occurred.

Skepticism notwithstanding, earthquake prediction research has continued within the seismology community in recent decades. Of particular note, several methods have been developed around the theory that, prior to a large earthquake, a region will experience an upsurge of moderate quakes surrounding the impending large event. The idea of distinct precursory patterns of regional earthquakes—for example, recall the Mogi doughnut, described by Kiyoo Mogi in 1969—prior to a big quake is not new. Some researchers have found (supposed) evidence for precursory quiescence; others have claimed a precursory upsurge in activity.

This type of approach falls within the umbrella of pattern-recognition techniques that seek to essentially read the tea leaves, looking for characteristic patterns of small or moderate quakes that tell us a big quake is on the way. Russian seismologist Vladimir Keilis-Borok has worked for decades on the so-called M8 algorithm, which relies on observed earthquake patterns to issue "TIP" (Time of Increased Probability) alarms. The full M8 algorithm is complicated, but most fundamentally TIPs are based on observed increases in regional seismic activity as the key harbinger of a future large quake.

The idea of regional activity increase derives a measure of support both from theoretical considerations and from observations. Seismologists have long been aware that the San Francisco area experienced significantly more felt and damaging quakes in the years prior to 1906 than in the decades after. And then, of course, there was Haicheng.

The M8 algorithm has been around long enough to have an established track record of hits versus misses. The method scored an apparent hit with the 1989 Loma Prieta ("World Series") earthquake south of San Francisco. Following the earthquake a number of researchers concluded that regional activity had increased prior to 1989. But it turns out to be surprisingly difficult to evaluate the overall success of any prediction method. In short, one has to show that a method is more successful than one could be by making predictions based simply on our general understanding of where and about how often large earthquakes occur. One could, for example, issue a series of three seemingly specific predictions for the following year: (1) A magnitude 5.0 or greater earthquake will strike within three hundred miles of Los Angeles; (2) a magnitude 6.5 or greater quake will strike within three hundred miles of Padang, Sumatra; and (3) a magnitude 6 or greater quake will strike within three hundred miles of Tokyo. On any given year these predictions are likely to do pretty well simply on the law of averages, given what we know about how often earthquakes strike these three areas.

LOMA PRIETA

Most seismologists do not believe the 1989 Loma Prieta earthquake was predicted per se. But it did break along, or immediately adjacent to, a segment of the San Andreas Fault that scientists had their eye on for decades. As summarized in a 1998 article by Ruth Harris, no fewer than twenty studies had identified this segment as ripe for a large earthquake. Many but not all of the studies were essentially based on seismic gap theory; that is, the fact that this segment of the fault experienced less slip than other parts of the San Andreas in the great 1906 San Francisco earthquake. The counter-argument (i.e., that no gap existed) was advanced by scientists who looked at geodetic surveying data before and after the earthquake, data that geophysicist Chris Scholz believes were never compelling due to limitations of the surveys. At a minimum it is fair to say that, as earthquakes go, Loma Prieta was more anticipated than many. (The fact that Loma Prieta apparently did not occur on the San Andreas proper, but rather an adjacent dipping fault, is, however, a minor wrinkle.) Gap theory does not lend itself to short-term prediction; nor is it clear that it lends itself to meaningful intermediate-term forecasting. Most scientists do, however, believe that it can point to segments of faults that have to break, some day.

To evaluate rigorously the success of a prediction method is thus a major research effort in its own right. In recent years seismologist Tom Jordan has worked to launch the Center for Earthquake Predictability, an international consortium effort aimed not at prediction per se, but at understanding predictability and evaluating prediction methods. Jordan and his team hope to formalize the evaluation process, not only of so-called alarm-based methods like M8, which predict earthquakes within specified time and space windows, but also of proposed forecast models that predict how many earthquakes are likely in a given region over a given time period.

The track record of the M8 method has already been evaluated in a number of studies. The method has been a moving target, as Keilis-Borok and his colleagues have continued to further develop and refine the methodology. By 2003 they had developed a more complicated method based on not only regional earthquake rates but also the identification of "earthquake chains." An earthquake chain is defined to be a series of earthquakes above a certain magnitude, within a certain period of time, with locations that define something like a line stretching over a certain distance on a map. The definition sounds vague: given any map of recent earthquakes, finding events that seem to align is about as meaningful as finding a cloud that looks like an elephant. But the definition of a chain is precise, specifying what is meant by "a certain magnitude," and such, and what exactly constitutes a chain pattern. Once the definition is specified, a chain is identified entirely objectively, based on established criteria.

Looking at patterns of small and moderate quakes prior to large earthquakes that occurred in California before 2000, Keilis-Borok and his team developed an algorithm that was very successful in "postdicting" the earthquakes. Postdicting, which basically means predicting earthquakes after the fact, might sound like shooting fish in a barrel. But there was more to the Keilis-Borok method than that. Their prediction scheme relied first on the identification of an earthquake chain. If a chain popped up, one then considered the earthquake activity in a region around the chain. If activity within that region had shown an increase over recent months or years, then an alarm is declared.

One can think of the scheme, or algorithm, as a computational thresher: once the machine is put together, one can throw in any catalog of earthquakes and the thresher will separate the meaningful precursory patterns from the chaff. The machine was constructed in the first place based on past earthquake catalogs. That is, it was constructed to successfully postdict large quakes that had already happened. But the fact that one can construct a successful thresher suggests that certain patterns do reliably precede large earthquakes. Or so one would think.

Predicting earthquakes after the fact might be a noteworthy achievement, but the point of earthquake prediction is to predict earthquakes before they happen. Keilis-Borok and his colleagues began making so-called forward predictions—that is, predictions of earthquakes that haven't yet happened—in 2003. At the time the predictions began they were not formally documented or systematically circulated within the community. The predictions specified the location to be a swath around the identified earthquake chain, a magnitude, and a nine-month time window. Keilis-Borok and his colleagues began to generate a buzz within the field with two apparently successful predictions: one in Japan, and one in central California that was fulfilled by the 2003 M6.5 San Simeon quake, which struck near the city of Paso Robles.

They generated even more buzz in early 2004 with the prediction of a magnitude 6.4 or greater earthquake in the southern California desert, some time before September 5 of that year. This prediction found its way out of the sheltered waters of academic discourse and into open, public discourse. As it happened the annual meeting of the Seismological Society of America (SSA) was held in Palm Springs in April of 2004. The schedule of talks for these meetings is always planned in January so that the program can be published prior to the meeting. When news of the Keilis-Borok prediction reached conference organizers, meeting organizers added a late-break slot to one session to give Keilis-Borok a chance to present his method and discuss the prediction. The talk was at five o'clock, in the largest of the three conference rooms.

Seismologists went into the room with varying degrees of interest and skepticism, the latter in some cases loudly voiced. But the room filled up, and stayed full as Keilis-Borok's talk stretched well beyond the fifteen minutes allotted for presentations, and as the question-and-answer session stretched for another half hour. Born in Moscow in 1921, and recognized in seismology for a number of seminal contributions early in his career, in his mid-eighties Keilis-Borok remained intellectual intensity personified. Some seismologists harbored suspicions about his continued fervor for prediction research. But whatever skepticism scientists brought into or took out of the meeting room in Palm Springs, the room stayed full until the audience ran out of questions. Keilis-Borok answered all of the questions with candor. Pressed on what physical mechanism could explain earthquake chains, he offered a few reasonable ideas but admitted that he didn't know the true answer. Indeed, it isn't necessary to understand why patterns occur to develop a successful method to predict earthquakes based on patterns. Pressed on how likely the predicted earthquake was, Keilis-Borok replied that the odds were very high, maybe 90 percent, "almost certain."

A number of journalists came to the SSA meeting. The prediction story, which had already garnered media attention, found fresh wind in its sails. By the summer of 2004 word of the prediction had spread in the desert region—Palm Springs and its environs—to the point that it became a topic of conversation over water coolers and gas pumps. Many people didn't have their facts entirely straight; one common misperception was that the quake was predicted to happen on, rather than before, September 5. But the man on the street knew that "scientists had predicted a big quake." And they were worried.

At the end of the day, of course, earthquake predictions hit or they miss. Keilis-Borok's prediction was a miss. If anything the southern California desert region remained stubbornly, perhaps unusually, quiet throughout the summer and early fall of 2004. The prediction window closed without a whimper, let alone a bang.

When some journalists came back for interviews after the prediction window closed, Keilis-Borok downplayed the significance of the single failed prediction. He noted, not unfairly, that the success of a prediction method could only be established by a track record of hits and misses. This one specific prediction, he told them, had only ever been expressed in terms of probabilities, and had never been a certainty. The predicted quake, he now told them, had only ever been perhaps 50 percent likely.

By the end of 2007 Keilis-Borok's group has amassed a track record of predictions. Following the two apparent hits in Japan and central California in 2003 the predictions have missed the mark far more often than not. Jeremy Zechar, working with thesis advisor Tom Jordan, evaluated the record using careful statistics. The two early hits notwithstanding, by the end of 2007 the method was looking like a statistical bust. For starters, one had to question whether or not to give the method credit for the two initial hits, these predictions having been brought to light only after the earthquakes had occurred. Were other predictions made quietly around the same time, predictions that ultimately failed? Without systematic documentation we have no answer to this. This question aside, of twenty subsequent predictions, only two could be considered hits. Of these, one was for a magnitude 5.5 or greater earthquake that Zechar calculated had been over 60 percent likely to happen by random chance. The second apparent hit was for a magnitude 7.2 or greater earthquake that was 15.9 percent likely to happen by random chance. But considering the overall track record of hits and misses, Zechar concluded that the Keilis-Borok method was no more successful than educated guessing—that is to say, making predictions based on the known average rates of moderate and large quakes in a given area.

One of Keilis-Borok's collaborators, Peter Shebalin, also evaluated the track record of the method, and arrived at different conclusions. According to Shebalin's calculations, the track record of the method was significantly if not overwhelmingly better than educated guessing. Clearly Shebalin's evaluation of statistics was different from Zechar's. For one thing, the prediction method had resulted in a series of almost continuous predictions for the southern California desert region, each new prediction associated with a newly identified earthquake chain. In Shebalin's calculations he considered this to be one failure, whereas Zechar pointed out that each prediction should be counted as a separate prediction and a separate failure.

So what went wrong? How could the thresher so successfully work on past earthquakes, with very few false alarms, and then go berserk with false alarms when set into motion looking forward? The answer lies in the dark and murky realm of statistics. When Keilis-Borok's group worked to identify patterns—in particular chains—that had formed before large quakes and not at other times, they had a lot of free parameters to play with. In mechanical terms, they put the thresher together with lots of knobs, all tuned to successfully separate wheat from chaff given known inputs and known outputs. Although it seemed like they had identified meaningful patterns, it turned out they had given themselves so many knobs to play with that they succeeded only in turning the knobs to explain past patterns. Put through its paces with real data, the thresher couldn't separate beans.

The 2004 prediction for southern California, coming on the heels of two apparently successful predictions, did attract genuine interest and curiosity within the seismological community. Many who filed into the meeting room in Palm Springs were confirmed earthquake prediction skeptics, but still they filed into the room, and they stayed in the room until the audience ran out of questions. Whatever curiosity or optimism they brought into the room did not dissipate immediately, but it did wane fast a few years later as the results of Zechar's careful analysis—as well as the track record of failed predictions—started to be known in the seismology community. Thus was the lesson hammered home anew, that apparent anomalies identified prior to past earthquakes, no matter how seemingly compelling, have a knack for unrolling themselves and crawling away just as one gets the mallet squared away. Thus were seismologists reminded all over again that earthquake prediction, like croquet in Wonderland, is a difficult game to play.

CHAPTER 12

Accelerating Moment Release

> When I see that I am wrong I change
> my mind. What do you do?
> —JOHN MAYNARD KEYNES

Although the U.S. seismology community remained generally pessimistic about earthquake prediction even a quarter-century after the Palmdale Bulge deflated, researchers besides Keilis-Borok have pursued research focused on the notion that regional activity increases prior to large earthquakes. In particular, while Keilis-Borok's group developed the M8 and later methods at UCLA, Charlie Sammis and his graduate student David Bowman pursued independent research at cross-town collegiate rival USC, working in collaboration with Geoffrey King in Paris. They focused specifically on testing the hypothesis that activity increases, known in seismological circles as the Accelerating Moment Release, or AMR, hypothesis. The M here stands for seismic moment, the quantity now used by seismologists to quantify the size of an earthquake. Introduced by seismologist Keiiti Aki in 1969, the moment of an earthquake is a measure based on the force required to overcome friction on a fault and the average slip, or movement, during an earthquake. In physics parlance, "moment" is similar to torque: the product of a force times a moment arm that is perpendicular to the direction of the force. A seismic moment is not the same thing because the slip direction is parallel with the direction of the force, but the units are the same. In effect, seismic moment reflects directly the overall size, and seismic energy release, of an earthquake. (Richter's magnitude scale, introduced in 1933, also classifies earthquakes based on overall size, but whereas magnitude units are arbitrary, seismic moment describes energy release in terms of well-known physical parameters.)

The AMR hypothesis dates back to the 1980s, when seismologists Charles Bufe and David Varnes first began to look at the patterns of increasing regional activity that had apparently preceded both the 1906 San Francisco earthquake and the 1989 Loma Prieta earthquake (fig. 12.1). In the late 1990s Sammis and Bowman began looking carefully at past moderate and large earthquakes in California, looking for evidence of similar patterns. By 2001 they showed that the AMR pattern had preceded a number of big earthquakes in California. Further, they showed that the bigger the eventual large quake, the bigger the region in which AMR was observed. The results seemed not only compelling but also conceptually appealing and reasonable. First, the work appeared to confirm the pattern that had been recognized before the 1906 and 1989 quakes in northern California. And in general terms, theories of damage mechanics predict that a system will experience a progressively increasing series of small failures as it approaches catastrophic failure.

Further development and testing of the AMR method became the centerpiece of Bowman's PhD thesis. In particular, he and his advisors worked to develop a theoretical basis for the observation. By 2000, work by Geoff King and others had established that stress changes following large earthquakes could play an important role in controlling the locations of aftershocks, and sometimes subsequent large earthquakes. To explain AMR, Bowman and his colleagues essentially turned these theories inside out. As regional stresses built up to a large earthquake, they reasoned, the region would respond to, essentially, the negative of the stress change that would occur in the large quake. The details are complicated, but basically predict that the rate of moderate quakes will increase in areas where stresses would decrease when the earthquake happened.

Details aside, Bowman's theory led to a testable hypothesis, which is what good scientific theories are supposed to do. Scientists can calculate the pattern of stress change that an earthquake will produce. For the most part, of course, earthquakes release stress, but they will also

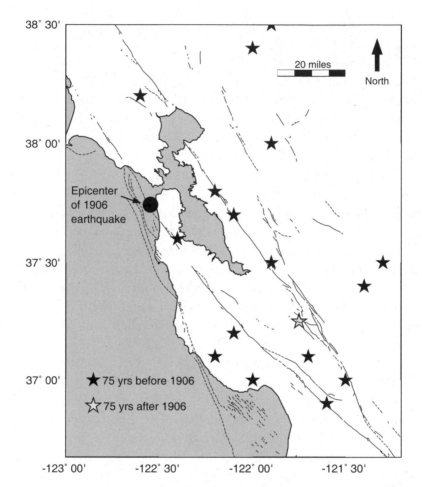

Figure 12.1. Map shows moderate earthquakes in the San Francisco Bay Area during the seventy-five years before the 1906 San Francisco earthquake (black stars) and moderate events during the seventy-five years after the 1906 earthquake (open star). Circle indicates inferred epicenter of the 1906 earthquake. (Image courtesy of Susan Hough, after work by Ross Stein and others.)

raise stress in certain areas; for example, if one segment of a fault breaks, it tends to load up the immediately adjacent fault segment. Thus, given an earthquake like the magnitude 7.3 Landers event in 1992, Bowman and his colleagues could calculate the stress change and then use it to define a region where they predicted AMR would be observed. They could then compare the AMR signal in the identified region with the signal in a simple circular or elliptical region around the earthquake. If their theory was right, the former signal should be stronger than the latter. Bowman did these calculations for recent large earthquakes in California. Again, the results seemed compelling.

Aware that subtle, sometimes fatal problems can be lurking inside apparently compelling results, Bowman and his advisors devised careful tests of the statistical significance of their results. They cooked up so-called synthetic catalogs, basically observed earthquake catalogs in which the times of all events were randomized, and showed that the AMR signals were highly unlikely to occur by random chance.

The AMR method does not lend itself to short-term prediction. From the beginning, scientists who worked on the method were aware that, looking forward, it was difficult to look at an accelerating signal and know exactly when the large quake would occur. Instead, the method appeared to offer hope that we could identify regions that were ripe for a large quake to occur on a time scale of decades, or perhaps years. Although well short of the specific earthquake predictions the public would like, such short-term forecasts, or intermediate-term predictions, would obviously be of enormous value. Bowman's refinements of the method appeared to offer further hope that one could eventually predict the location as well as the magnitude of the impending quake.

By 2000 Bowman and his colleagues had done considerable work looking at patterns before past earthquakes and started to move toward forward predictions. They considered the segments of the San Andreas Fault that had broken in 1857 and 1906. They found no evidence for an AMR signal pointing to an imminent repeat of the 1906 earthquake but did find evidence for increasing activity around the 1857 rupture zone. Aware that their research was both a work in progress and a potential hot potato, they opted to write the paper in French and publish it quietly in the Proceedings of the National Academy of France.

As Bowman and his colleagues pressed forward to develop and test his method, another seismologist, Andy Michael, began to ruminate on the results he had seen presented in journals and at meetings. Drawing on previous experience analyzing the statistics of a prediction method based on very-low-frequency (VLF) radio waves, Michael began to wonder if subtle and fatal biases might in fact be lurking in Bowman's observational analysis. In particular, Michael suspected that Bowman's group had also given themselves so many knobs to turn that they could find AMR signals where none existed, for example by adjusting the size of the region or the time period they used to look for AMR. Bowman's own statistical tests seemed to say otherwise: that the results were both significant and robust.

But again, where earthquake patterns are concerned, reading the tea leaves is a tricky business. It is often a lot easier to get apparently compelling results than to prove that they are indeed meaningful. In this case, Michael suspected a flaw in the tests that Bowman had devised. Bowman had compared his results with results from catalogs that had the same numbers of earthquakes of various magnitudes as the real earthquake catalog, but in which earthquake times were random. Michael suspected Bowman's tests were not really putting the method through its paces because the tests did not take into account the propensity of earthquakes to cluster.

For several years Michael watched from the sidelines as Bowman and his colleagues continued to present apparently promising results. He had a number of opportunities to pass along his concerns but did not feel they were being addressed fully. Scientists often hesitate to dive into rigorous refutation of their colleagues' work for the simple reason that it can be a time-consuming endeavor, and not as gratifying as pursuing one's own research interests. But by the end of 2005 circumstances conspired to ratchet up Michael's concerns. First, attending a meeting of a working group that was about to undertake a major study to assess the statewide probabilities of earthquakes in California, Michael grew concerned when the leader of the working group pointed to AMR as a potential way to improve short- and intermediate-term earthquake forecasts. Were this done, AMR would leap out of sheltered academic waters and into the real world where decisions involve real dollars and cents.

Then at the December 2005 annual meeting of the American Geophysical Union it seemed to Michael that AMR was suddenly in the air. A year after the 2004 Sumatra earthquake, something akin to a bandwagon seemed to be underway. Talk after talk focused on AMR signals that had supposedly been seen before large earthquakes, including Sumatra, and signals that pointed to future large earthquakes. At this point Michael decided it was time for action. During the meeting he arranged an informal meeting with two colleagues: Jeanne Hardebeck, who had previously worked on stress-change theories, and Karen Felzer, who had established herself as a leading expert in earthquake statistics. Felzer hails from a hardy breed of earthquake statisticians who have a knack for occasionally annoying their colleagues by holding their feet to the statistical fire, insisting that hypotheses be formulated carefully and tested rigorously.

The discussion quickly grew into a collaboration. Michael and his colleagues devised what they considered to be a rigorous test of the AMR hypothesis, repeating Bowman's tests using not a catalog with random event times but a randomized catalog that included the same amount of clustering that is found in real earthquake catalogs. These tests led to quite different conclusions. Michael's team showed that it was in fact easy to turn the knobs to find AMR where none existed. Not only that, they could tune the knobs to find apparently compelling observations of DMR—decelerating moment release. These results suggested that Bowman's group had found AMR signals only because, essentially, their approach gave them enough latitude in picking the size of the region and the length of the time window prior to the earthquake. It was in effect the same kind of knob-tuning exercise that Keilis-Borok's group had done. The analysis that had seemed so careful, the statistical tests that seemed so reasonable, the results that had seemed so compelling—in the end, all the product of biases so subtle it took three top seismologists years of careful work to get to the bottom of the story.

Science is often a slow process; any given scientific paper can be years in the making. Typically, scientific meetings are where the action is—where new results are presented, discussed, and sometimes, debated. Michael and his colleagues first presented their work on the AMR hypothesis at meetings in 2006.

By this time Bowman had invested many years of work—the research in his PhD thesis as well as work in subsequent years—on observations of and the theoretical framework for the AMR hypothesis. As he moved on to become a professor of geology at California State University at Fullerton he continued what he was convinced was a promising and exciting avenue of research. When Michael and his colleagues began to show up at meetings with their negative findings, the dialog between the two research groups was cordial but could fairly be described as tense. Bowman described the statistical tests that he and his colleagues had done to demonstrate that AMR was real; Michael and his colleagues described the shortcomings of these tests.

The challenge inspired Bowman and his colleagues to revisit their own tests. By 2007 they had also acquired several years of experience looking for AMR patterns that would signal the occurrence of future large earthquakes. Via these experiments, in particular via attempts to identify AMR patterns that could be considered robust, they began to realize just how slippery the apparent patterns could be.

On any given day, any given scientist would rather be right than wrong. But science will always proceed with two steps forward, one step back—not simply because scientists are fallible individuals who make mistakes, but because the business of science is, at the heart of it all, the business of hypothesis testing. Hypotheses are developed, tested, sometimes proved, and sometimes refuted. Refutation might not be as gratifying as proof, but it is every bit as critical. Not every hypothesis will be right. As scientists push into the unknown some paths will lead us forward, and some, however promising they might have looked at the outset, will turn out to be dead ends. One can't hope to push boldly into the frontier of knowledge and pick the right path every time. One does however have to hope that, by the process of critical self- and/or collective examination, researchers will recognize errant paths. Recognizing a path as a dead end forces researchers, and the field as a whole, to regroup—to take a step back, to think critically about assumptions. This in turn can lead to new ideas, new hypotheses, new paths. Science goes wrong when scientists refuse to recognize evidence that is staring them in the face, telling them they are on the wrong path.

By late 2007 when Bowman described the AMR method in lectures to his students at Cal-State Fullerton, he titled the lecture "The Life and Death of a Prediction Method." He further joked the lecture might be subtitled, "The Life and Death of a Scientific Career." Of course Bowman's work on the AMR hypothesis was no such thing. A scientific career can be proclaimed dead not when a researcher himself realizes that a midcourse correction is in order, but when a researcher fails to admit that he or she is on a wrong path. Doggedly pursuing failed theories, tenaciously clinging to flawed observations, such are paths to career perdition.

Bowman's early career might not have played out the way he thought it would; the way he hoped it would. Apart from having seen his hypothesis refuted, he felt a measure of responsibility and regret for the students he had worked with over the years. As a mature, established, and by his own account emotionally sturdy researcher, Bowman understands that science is sometimes a contact sport. As a faculty member at a state university where many students are the first from their families to pursue higher education, he also knows that the students who worked with him on AMR research were, as a rule, rather less sturdy. He is left with particular pangs of regret at the extent to which they were subjected to harshly voiced criticisms—what Bowman regards as a hazing process that assertive students and young scientists might survive, but diffident students and young scientists sometimes do not.

The development, testing, and ultimate debunking of the AMR method was not the neat, clean, orderly business that many imagine research to be. The process of refutation can be harsh; students and young scientists are sometimes caught in the crossfire. But this is the face of science.

CHAPTER 13

On the Fringe

> What ails them is exaggerated ego plus imperfect or
> ineffective education, so that they have not absorbed one
> of the fundamental rules of science—self-criticism. Their
> wish for attention distorts their perception of facts, and
> sometimes leads them on into actual lying.
>
> —CHARLES RICHTER, unpublished memo, 1976

Seismology has the distinction, if one can call it that, of being not only a science that a lot of people care about, but also a science that a lot of people think they can do. In particular, one needs no test tubes, no clinical trials, no computers ... pretty much nothing but a soap box to step forward with an earthquake prediction. It is a game that, literally, anyone can play.

It is moreover a game that people have been playing for a very long time. The persistent notion of earthquake weather dates back at least as far as the fourth century BC, when Aristotle proposed that earthquakes were caused by subterranean winds. From the time that seismology emerged as a modern field of scientific inquiry in the eighteenth century, scientists speculated about associations between earthquakes and weather (among other things). Not long after a fearsome series of earthquakes struck the North American midcontinent over the winter of 1811/12, stories began to circulate that Shawnee leader Tecumseh had predicted—prophesied, even—the events. That the historical record contradicts the story has not stood in the way of a good yarn. A 1990 circular published by none other than the U.S. Geological Survey boasts the title, "Tecumseh's Prophecy: Preparing

Figure 13.1. Charles Richter. (Photograph courtesy of California Institute of Seismology Seismological Laboratory.)

for the Next New Madrid Earthquake, a Plan for an Intensified Study of the New Madrid Seismic Zone."

I will return to New Madrid in the next chapter, but simply note here that, looking back through the annals of amateur earthquake predictions, predictions and predictors start to seem as interchangeable as widgets. The papers of the late Charles Richter illustrate the point (fig. 13.1). In the halcyon times before the advent of the World Wide Web, scientists could opt for quiet lives of productive anonymity, recognized within but rarely outside of their professional communities. Throughout much of the twentieth century Richter was among the small handful of earthquake scientists whose names appeared regularly in the media. The public came to know Charles Richter largely because of the scale that bears his name, but also because he talked to the public at a time when the same could not be said of most seismologists. As a consequence of these efforts, Richter found himself on the receiving end of torrents of mail—letters from people who were afraid of earthquakes, letters from people who were interested in earthquakes, letters from people who were convinced that they could predict earthquakes.

In the mid-twentieth century as now, amateur earthquake prediction schemes fall into several basic classes. Triggering by tidal forces is a common theme, leading to predictions that large quakes will occur at times of full or new moons, and in particular when the solar and lunar forces conspire to produce seasonal high tides. In *The Jupiter Effect*, a well-known book first published in 1974 at the height of the earthquake prediction heyday, authors John Gribbin and Stephen Plagemann argue that a 1982 alignment of the planets "might well trigger a California earthquake far worse than the San Francisco catastrophe of 1906." The book generated enough waves to be the subject of a (highly critical) *Los Angeles Times* article in September of 1974. Tidal forces do generate not only ocean tides but also tides, and therefore stresses, in the solid earth. But these stresses are extremely small, and careful research has shown them to cause at most a very small modulation of earthquake rates.

Other predictions hark back to Aristotle, predicting earthquakes based on some aspect of the weather—temperature, rain, or atmospheric pressure. Richter himself once wrote that he had always noticed a slight increase in the number of small earthquakes in southern California in late fall, shortly after the first rains, which he attributed to the "movement of air masses."

Some individuals are convinced that signs of impending large earthquakes are somehow "in the ether," by virtue of either precursory electromagnetic signals or vibes that can start to sound like George-Lucasesque disturbances in The Force. Some individuals are convinced that they are sensitive to these signals. The manifestations of these sensitivities vary between individuals: some have aches and pains, others hear noises, still others a vague sense of unease. Sensitivities are sometimes cast in spiritual terms: the voice of God, the angelic apparition, the warning from beyond the grave.

In a book published in 1978, *We Are the Earthquake Generation*, author Jeffrey Goodman, who had a PhD in archaeology, described the remarkable consensus among psychics, past and present, that 1990–2000 would be the "Decade of Cataclysm." From Nostradamus in the sixteenth century to America's Edgar Cayce in the early twentieth century, self-professed psychics were unified in the belief that the planet would undergo a period of major upheaval prior to the new millennium. Goodman interviewed a number of notorious psychics, all of whom agreed that the world was about to be "shaken from top to bottom." Among the specific predictions:

- Major sections of the western United States fall into the sea as the coastline moves eastward in a series of violent surges;
- major disturbances (earthquakes and subsidence) on both the east and west coasts of the United States;
- portions of northern Europe break away or sink;
- most of Japan goes under the ocean.

And so forth. After the first edition of the book was published in 1978, the Berkley Publishing Corporation asked Goodman to put together a second edition, adding new material to identify the ten most dangerous cities in the United States.

Chapter 10 of Goodman's book, "Earthquake Prediction: Consult Your Local Cockroach; or, How a Cockroach Could Save Your Life!" explored the enduring notion that animals have the ability to sense impending large quakes. To this day the idea remains deeply ingrained among the general population. After all, as discussed earlier, even the trained seismologist cannot explain or entirely dismiss what appear to be reliable accounts of odd animal behaviors prior to the 1975 Haicheng quake, although neither can we rule out the most straightforward explanation, namely that animals responded to the foreshocks. Although the handful of controlled investigations has never demonstrated any correlation between animal behavior and earthquakes, the belief lives on. And as with the case of other proposed precursors, it remains within the realm of possibility that the belief lives on in part for good reason, that is, that animals do in some cases react to anomalies, for example gas release, that are related to earthquakes, but that animal behavior, like the anomalies themselves, are of no practical use for prediction.

Some amateur predictors pursue their avocation more seriously than others. Self-described futurist Gordon-Michael Scallion for years published the regular "Earth Changes Report," the paper newsletter in recent years replaced by a polished Web site. Scallion claims to have received a series of "out-of-body" visions in 1979, revealing what the world would be like after the turn of the century. Scallion spent the next three years putting together a map of the world with the dramatic changes he had foreseen, changes that included massive inundation of coastal regions as well as along major river valleys such as the Mississippi. Three decades later, Scallion and his followers point to the remarkable prophecies that (or so they claim) went unheeded by most of the world: increasingly erratic weather, melting of the polar ice caps, increasingly frequent and severe tornadoes.

In June of 1992 a large earthquake tore across a sparsely populated desert region in southern California, generating minimal damage by virtue of its remote location but much excitement in the earthquake science community. At magnitude 7.3, the Landers earthquake was the largest event in southern California in forty years, and the first big quake in the region since the USGS first got into the earthquake monitoring business in the late 1970s. We learned a lot from Landers, in particular about how large and small earthquakes disturb the crust and sometimes trigger other large and small earthquakes. The Landers earthquake had a more muted impact on the public psyche than the 1971 Sylmar quake; the former, while larger, had not been an urban disaster.

Not long after the Landers earthquake, copies of the "Earth Changes Report" began to show up at the U.S. Geological Survey office in Pasadena. My colleagues and I learned that the Landers quake, and the magnitude 6.1 Joshua Tree quake that had occurred two months earlier, were part of a cataclysmic sequence that Scallion had predicted. In fact they were only the beginning. Larger earthquakes were predicted in the fall of 1992 and beyond, a series of catastrophic events akin to the doomsday scenario that later played out in spectacularly laughing fashion in the made-for-television movie *10.5*. But Scallion's reports were serious. No more fun and games, people; California was going to fall into the ocean. Snap up that oceanfront property in Arizona before it's too late!

Scallion's reports didn't just predict earthquakes in southern California; they predicted earthquakes in other active parts of the world as well as various other global catastrophes including storms and political unrest. The months wore on and the reports, each with many doomsday predictions, piled up. Fall of 1992 came and went without further excitement in southern California. After a while one began to understand how the game was played. The "Earth Changes Reports" would comment on past predictions, making hay of events such as Landers that fit—perhaps with the help of a shoehorn—one of the predicted scenarios. The reports would also acknowledge that some small part of some specific prediction had not happened. Missing in the reckoning was any acknowledgment that doomsday had not, in fact, arrived; the predicted megaquake had not struck; California had not fallen into the ocean. Indeed, these predictions continued, the timing of the apocalypse shifting steadily forward in time. Eventually a big quake, larger than Landers, is going to strike southern California. We don't know when it will be but we can be sure of one thing: Gordon-Michael Scallion, if he's still around, will have predicted it.

Amateur earthquake predictors might start to feel interchangeable, but they are a varied breed. If Scallion represents one end of the spectrum, Zhonghao Shou perhaps represents the other. Shou, a retired chemist from mainland China, has for years believed that the key to earthquake prediction is in the clouds. Shou has identified certain cloud formations, what he refers to as earthquake clouds, that he claims are different from any of the standard cloud formations generated from natural atmospheric conditions. His avocation traces back to a photo he snapped the day before the 1994 Northridge earthquake, a wispy, linear, near-vertical cloud in the sky above the San Fernando Valley. Since that time he has scoured available data banks for signs of other earthquake clouds. His endeavors earned him the nickname Cloud Man, a name he was happy to embrace.

Shou has sought out members of the earthquake science community. He has done a lot of talking; the conversations tend to be one-sided, but Shou has teamed up with collaborators who have also listened. For years he regularly sent specific prediction to the USGS office in Pasadena, where public affairs officer Linda Curtis filed them. He has discussed a theoretical basis for earthquake clouds, ideas regarding heat and/or vapor release that are not unlike ideas discussed by at least some scientists in the mainstream community. By 2004 he set out to evaluate his track record statistically. To date scientists have not found the results of these tests compelling. At issue is the familiar question of whether Shou's predictions have done better than random chance, given that his method identifies fairly long prediction windows in areas where, as a rule, large earthquakes are not uncommon. Still, among the ranks of amateur predictors, Shou has distinguished himself for pursuing his avocation with more rigor than most. In this Shou is unusual but not unique. Another team of amateur predictors, Petra Challus and Don Eck, have teamed up to make predictions based largely if not entirely on observed short-term earthquake rates. In her overall views on earthquake prediction, Challus is not always on the same page as the mainstream earthquake science community. But the approach used to make predictions for the Quake Central Forecasting Web site appears to be similar to, if less rigorous than, earthquake-rate-based methods now under development by mainstream researchers. Whether predictions on the Web site will some day depart from this sort of approach remains to be seen. At least initially, Challus and Eck appear to be playing fair. To the seismologically trained eye they are predicting small to moderate quakes that are likely to happen given background rates of earthquakes. Another Web site, run by Brian Vanderkolk, also issues predictions based on simple and clearly stated rules. Vanderkolk points to the apparent success of his method as an argument against touted successes on the part of other would-be predictors. The above efforts reveal that the "fringe element" is not only a continuum, but one that at least occasionally edges closer to the mainstream scientific community than scientists might care to admit.

One could fill an entire book with amateur prediction efforts: the ideas, the personalities, the reasons why theories and/or prediction track records are problematic.

One can however note that, for all of the predictions and all the predictors, a reliable prediction scheme has yet to emerge, from any quarter. Scientists tend to think in terms of physics-based predictions; we believe that reliable predictions, if they are ever possible, will be tied to an understanding of the how earthquakes happen. We tend to dismiss prediction schemes that are based on theories we know, or at least have good reason to believe, are not credible. As earlier chapters have discussed, of particular note is that slippery assumption that underlies so many prediction schemes, namely that stress builds rapidly prior to a large earthquake, and causes significant changes in the earth. If the earth did behave this way, accelerating stress (think of it as pressure) would give rise to accelerating strain (warping in response to pressure). Yet by now scientists have recorded extremely precise measurements of strain in the crust prior to a number of nearby earthquakes and we have seen nothing—no signals that are big enough to stand above the noise. These results do not prove that precursory strain signals don't exist; however, they do tell us that, if such signals exist, they are very small.

Some prediction schemes, in particular those based on pattern-recognition, sidestep the question of physics, at least at first, instead focusing on the quest to find patterns in the tea leaves. In the end prediction schemes prove their salt not by the soundness of the underlying physics but by their track record of success. Had Keilis-Borok racked up a series of hits with his predictions, skepticism would have fallen by the wayside.

Amateur scientists level various sorts of accusations at the mainstream scientific community, variations on a drumbeat theme that we are hegemonical, closed-minded, unwilling to acknowledge or accept breakthroughs that come from outside the ranks. But, again, the proof is in the pudding. If a self-professed holy man (futurist, whatever) were to rack up a series of truly successful predictions based on visions (visitations, whatever), the world would take note. The earth science community would take note. Without question, the insurance industry, for whom predictability is nothing less than bread and butter, would take note.

The mainstream community is quick to dismiss amateur prediction efforts because few if any come close to the bar established for modern scientific inquiry. Few if any are willing, or have the wherewithal, to turn their results inside out to see if they stand up to rigorous statistical analysis. Few if any are willing to even own up to their track record of hits and misses, let alone analyze their record with rigorous statistics.

One can consider the question: what impels amateur predictors? Having been on the receiving end of amateur predictions for decades, Charles Richter had his own answer. "A few such persons are mentally unbalanced," he wrote of amateur predictors, "but most of them are sane—at least in the clinical or legal sense, since they are not dangerous, and are not running around with bombs or guns. What ails them is exaggerated ego plus imperfect or ineffective education, so that they have not absorbed one of the fundamental rules of science—self-criticism. Their wish for attention distorts their perception of the facts, and sometimes leads them on into actual lying."

We tend to ascribe other motivations to members of the mainstream scientific community who develop prediction methods, or promote prediction programs. In his 1976 memo, written during the earthquake prediction heyday, Richter allowed that "occasionally a professional man who has a good reputation in other fields is responsible for erroneous statements about earthquake occurrence and earthquake prediction. Even good geologists have been known to fall into such errors." Richter might (or might not) have been understanding that Keilis-Borok's thresher went awry for understandable and forgivable reasons.

As a community we tend to look back at the 1970s with a measure of sympathy for the genuine optimism on the part of many top scientists and a measure of appreciation for those who parlayed public interest into effective science and hazard mitigation programs. Whether less altruistic motives were lurking beneath the surface, who is to say? Certainly scientific leadership is not without rewards.

Among scientists there is a presumption so deeply rooted as to be not only beyond dispute but also beyond conscious consideration: a scientist might think that another scientist is dead wrong in his interpretations, but never that he is outright dishonest. Splashy cases of scientific fraud rock the scientific community to its core because they challenge the very heart of our belief system, namely that scientists aren't always right, but they are always honest actors. Viewing with dispassion certain chapters in the annals of earthquake prediction, one might conclude that even among the ranks of mainstream scientists there sometimes seems to be a vanishingly fine distinction between genuine optimism and, if not charlatanism, at least opportunism.

The mainstream earth science community looks at the amateur earthquake prediction community with somewhat different presumptions. If shading the truth serves to launch worthwhile programs, we can live with a bit of that. If shading the truth serves to promote what we see as myths and false hopes, or blatant self-aggrandizement, we are rather less tolerant.

At the end of the day the mainstream community is left with two hard questions. First, do we hold ourselves to the high standards of intellectual integrity and critical examination that we are so quick to pull out to club amateurs who wander onto our turf? And, second, are we too quick to dismiss ideas that are out of the mainstream and, if unproven, are worthy of consideration?

Looking back at earthquake prediction research in the mainstream community, the answer to the first question has to be, "sometimes yes, sometimes no." Ideally, individuals who pursue prediction research will

themselves undertake critical examination of intriguing results—
maybe, as in Bowman's case, with a nudge from colleagues. This is the
way science is supposed to work; it isn't always the way that science
does work.

When individual researchers remain unbowed in the face of compelling criticism, eventually the scientific field will identify and prune
aberrant and unproductive limbs. But when they are fueled by the
fervor of what lifelong prediction skeptic Robert Geller calls faith-
based science, errant limbs can last a long time. Well intentioned or
otherwise, such research fuels false hopes but, worse, can soak up scarce
resources that could be spent more fruitfully elsewhere to mitigate
earthquake risk. Following the devastating magnitude 7.6 Kashmir
earthquake of 2005 global resources were suddenly available for hazard
mitigation in Pakistan, a country that faces an enormous earthquake
hazard with an enormously vulnerable building stock and infrastructure. In southern Pakistan, in and around the megacity of Karachi, we
lack the most basic information to identify active faults and quantify
the expected rate of future damaging earthquakes. Given what little
we know, it is possible that hazard in Karachi is comparable to hazard
in Los Angeles. It is also possible that hazard in Karachi is comparable
to hazard in eastern Nevada—which is to say, higher than some areas
but a lot lower than others. To decide what building codes are appropriate for Karachi, basic science is needed: field investigations to identify and investigate active faults, GPS measurements to find out how
fast stress is building on various faults, monitoring of small earthquakes
to help illuminate active faults in the crust. Money for these sorts of
studies has always been hard to come by. When, as is typically the case,
the 2005 disaster shook loose some resources, very little of that bounty
found its way to investigations that would address critical basic questions, yet some of the money went toward consultations with Keilis-
Borok's group on earthquake prediction.

Returning to the second question, one also has to consider whether
or not "fringe" methods might have some degree of underlying validity. The answer here is yes. As noted, scientists are justifiably skeptical
about methods that are contradicted by sound theory and/or observa-

tion. But other ideas remain plausible. Large weather fronts generate significant changes in atmospheric pressure; we now know that small pressure changes generated by earthquakes can sometimes trigger other earthquakes. A long line of serious scientists have given the matter consideration, including Robert Mallet, one of the founding fathers of the field of seismology. Mallet looked at observatory records prior to the great Neapolitan earthquake of 1857 and found that the mean barometric pressure was 10.76" above the mean during the same month over the four previous years, whereas the rainfall was significantly lower than average. Mallet's interpretation focused on the paucity of rain rather than the pressure change. Believing that earthquakes occurred as a result of disturbances in underground cavities, the mechanism that he proposed was, inevitably, total hogwash. But in his seminal 1862 publication—arguably the first comprehensive scientific report on an important earthquake—he noted that more work was needed to document a connection between atmospheric conditions and earthquakes. He added, "I think we may consider it highly probable that future and more extended observation in volcanic and seismic countries will establish this connection between meteorology and seismology."

Similarly, it is beyond dispute that tidal forces generate cyclical stresses in the earth. Can we say to a certainty that it was purely by coincidence that the monster M9.3 Sumatra quake of 2004 occurred during a bright full moon? In fact we can't. The theory of plate tectonics explains how and why stresses built up along the fault that produced the earthquake, but no known theory explains why the quake struck minutes before eight o'clock on the morning of December 26. Maybe tidal forces did have something to do with it. Once again it is critical to distinguish between the quest to develop reliable earthquake prediction and the quest to understand the physics of earthquakes. This is, tidal stresses are cyclic, and influence the entire planet. One clearly cannot hope to develop reliable, specific earthquake prediction methods based on such signals.

For all of the advances in earthquake science since Mallet's time, there is a lot about earthquakes that we still don't know, including perhaps the most basic question of all, why earthquakes start. Answers have been proposed within the mainstream seismology community; theories and models point to the underlying physics of earthquake nucleation. Some of these ideas might be right. It is also possible that the ultimate answer will come from what we now regard as the scientific fringe. On the one hand, as long as we don't know what we don't know, it remains incumbent on the mainstream seismology community to stay open-minded about ideas we don't much like. On the other hand, without hard, cold, statistical and scientific rigor, ideas alone will never move us forward. As the bumper sticker says, you don't want to be so open-minded that your brains fall out.

CHAPTER 14

Complicity

An incensed Charles Richter called this
prediction "the most terrific explosion of
pedigreed bunk" he'd ever seen in the earthquake
field, but reputable earthquake researchers were
not entirely innocent of fanning the fears.

—Carl-Henry Geschwind

*M*ost pseudo-science predictions languish in benign obscurity. Occasionally, however, one makes waves the conventional media, and reputable scientists are called to weigh in. The prediction that Richter described as "pedigreed bunk" was made in late 1968 by Elizabeth Stern, a self-proclaimed clairvoyant who had had a vision that California would be destroyed by a giant earthquake by April of 1969. As predictions go it was run of the mill. But this one generated a buzz, and the buzz swelled to a roar when a physics professor at the University of Michigan was quoted in a press release saying that within twenty years San Francisco would be struck by a great earthquake.

The prevailing winds in the larger earthquake science community, at that time pushing strongly toward the launch of the National Earthquake Hazards Reduction Program (NEHRP), also fueled the flames. Although most reputable scientists resoundingly dismissed the prediction, others spoke of signs that San Francisco faced imminent peril. In a March 1969 congressional hearing on the proposed new federal program, U.S. Geological Survey director William Pecora stated that a repeat of the great 1906 San Francisco earthquake would definitely occur by the end of the century, and probably by 1980. That same month scientists at the U.S. Geological Survey expressed concern that

Figure 14.1a. A sidewalk in Hollister, California, has been warped by slow creep on the Calavares fault. (Photograph by Susan Hough.)

accelerated creep along the San Andreas Fault near the town of Hollister (figs. 14.1a and 14.1b) might presage a major quake.

By the spring of 1969 talk of earthquakes was all over the airwaves in the San Francisco Bay Area. Many rushed to buy earthquake insurance; some went so far as to leave the state.

On April 28, 1969 California was rocked by an earthquake strong enough to cause tall buildings to sway. It was not, however, the doomsday event of Elizabeth Stern's dreams. The quake was centered in a sparsely populated part of the southern California desert, with a magnitude of 5.7. Eventually, inevitably, the Bay Area settled down and got back to business. But the flames were stoked. In March of 1969 California state senator Alfred Alquist decided to sponsor a new Bay Area earthquake commission. The proposed commission quickly expanded to have statewide scope and came into being that August as the Joint

Figure 14.1b. A concrete culvert at the DeRose winery in Hollister, California, has been offset by slow creep on the San Andreas Fault. Geophysicist Morgan Page stands with one foot on the North American Plate and one foot on the Pacific Plate. (Photograph by Susan Hough.)

Committee on Seismic Safety. On a national stage the push for NEHRP rolled on.

Science and pseudo-science collided again two decades later. This story ends in a different part of the United States, but begins in California on the evening of October 17, 1989, when an earthquake occurred along the San Andreas Fault in a sparsely populated part of the Santa Cruz Mountains south of San Francisco. After much debate most scientists conclude that the Loma Prieta earthquake was not ac-

Figure 14.2. Damage to the double-deck Nimitz Freeway in Oakland, California, from the 1989 Loma Prieta, California, earthquake. (Photograph by Susan Hough.)

tually on the San Andreas Fault, but rather on a different, nearby fault. Whatever fault it was, the impact of this earthquake, in societal as well as scientific circles, belied its modest magnitude. In the Marina district in San Francisco, expensive houses collapsed as the ground beneath them—artificial fill dumped along the edges of the bay—shifted and gave way. Across the bay a section of the double-deck Nimitz freeway suffered the same fate, for similar reasons (fig. 14.2). (The freeway was not built on artificial fill but rather on the soupy natural sediments, known as Bay mud, found around the fringes of San Francisco Bay.) A nearby section of the Bay Bridge also collapsed. Closer to the epicenter, downtown Santa Cruz took a heavy hit, as did working-class neighborhoods in the central valley town of Watsonville. The news trucks did not rush to Watsonville, although, while more expensive structures were later rebuilt, the quake took a heavy, lasting toll on the town's already scarce affordable housing.

Dramatic images of quake damage were beamed almost instanta-
neously to millions of homes across the country, in many of which an
audience had gathered to watch the World Series game scheduled to
play in, of all places, San Francisco. The game was, many feel, a stroke
of remarkable good fortune. With the game starting at five o'clock
many area residents went home from work early that day, putting most
of them in relatively safe single-family wood-frame homes, instead of
on area freeways, when the quake struck. One also has to remember
that, for all of the drama, Loma Prieta was a relatively modest quake at
magnitude 6.9, and it was centered well south of the urban centers
around the bay. The final death toll—always surprisingly difficult to
tally—stood at sixty-two.

The impact on the national psyche, however, went far beyond the
body count; beyond the $6 billion damage tally. For the first time since
1971 Americans saw an urban earthquake disaster in their own country,
a fairly modest one, but a disaster nonetheless. Many experienced the
earthquake firsthand, at least indirectly. Baseball fans around the country had just tuned into game three of the World Series between the
Oakland Athletics and the San Francisco Giants when San Francisco's
Candlestick Park sprang to life. Television announcer Al Michaels
managed to blurt out, "There's an earth . . ." before screens around the
country went dead.

By 1989 the NEHRP program was over a decade old. The program
could claim many successes by that point, earthquake prediction, of
course, not among them. As the program chugged along, increases in
funding did not keep pace with inflation. Worse, agencies like the
USGS found themselves with increasingly senior—increasingly expensive—work forces, running monitoring and other programs that
involved increasingly sophisticated—increasingly expensive—technol-
ogies. Short of another coordinated effort to launch a bold new program, earthquake scientists realized that earthquake disasters repre-
sented their one real hope to counteract the inexorable erosion of
support for programs.

The Loma Prieta earthquake did have an impact on politicians and
other policy makers, and did result in a boost to NEHRP funding. It

also raised awareness of earthquake hazard on a national stage, creating what we sometimes refer to as a teachable moment. Following any earthquake disaster, news stories follow a predictable course: initial stories, in locations both near to and far from the disaster, focus on the immediate story, in particular the impact on life, limb, and property. In short order, the news media in locations away from the earthquake inevitably ask the question, can it happen here?

In the U.S. heartland, a flurry of such stories began to appear within days of the earthquake. On October 18, 1989, just one day after Loma Prieta, readers of the Chillicothe, Missouri, *Constitution-Tribune* read that "the major earthquake that hit northern California could be a sneak preview for residents along the New Madrid fault." The next day, readers of the Springfield, Illinois, *Daily Herald* were informed that "the chances are 50-50 than an earthquake as bad as Tuesday's in California could strike Illinois within 10 years and spread destruction throughout the Midwest." "It's not a question of 'if.' It's a question of 'when,'" a high-ranking official with the Illinois State Geological Survey was quoted as saying.

These stories, and the concern, did not come out of nowhere. To understand the stage that was set in the aftermath of the Loma Prieta earthquake one must first step back in time. From seismology's inception as a modern field of scientific inquiry in the late nineteenth century, scientists were aware that one of the most portentous earthquake sequences in the United States had struck the midcontinent over the winter of 1811–12. The New Madrid sequence began without warning in the wee hours of the morning on December 16, 1811, with an earthquake strong enough to shake the bejeezus out of settlers on the frontier and to send powerful waves rippling across much of the country (fig. 14.3). The quake was centered in the boot heel of Missouri, some ninety kilometers (fifty-five miles) north of Memphis, Tennessee. (The modern city of Memphis was not founded until 1820; only Fort Pickering existed in the location at the time of the earthquakes.) The earthquake did not, as some who should know better still like to claim, ring church bells in Boston. Scouring Boston-area newspapers one finds no mention that the quake was even felt in the city. But waves

Figure 14.3. The 1811–1812 New Madrid earthquake sequence included at least four events large enough to cause liquefaction in the Mississippi River valley, including countless so-called sand blows. In this photograph taken about 100 years after the earthquake, sand blows can still be clearly seen. Many of these features remain apparent even today. (USGS photograph. Photograph by Myron L. Fuller.)

did plenty else: brick buildings cracked in St. Louis, chimneys toppled in Louisville, sleepers were awakened in Richmond, Virginia, and bells in church steeples in Charleston, South Carolina, were set ringing. Charleston is nearly one thousand kilometers (six hundred miles) as the crow flies from New Madrid. (One suspects but cannot prove that the enduring errant lore about church bells in Boston might have resulted from an early confusion between Charleston, South Carolina, and the Charlestown District of Boston.)

Around dawn that same morning another large quake struck, not quite as big as the first, but still large enough to be widely felt. An energetic series of aftershocks followed. Settlers close to the New Madrid region describe the ground as having been in an almost constant state of unrest. In Louisville, Kentucky, about 420 kilometers (250 miles) away, engineer Jared Brooks set out to record every earthquake large enough to be felt, and to rank quakes by relative severity. He went so far as to set up a number of pendulums that would swing in response to very gentle shaking. Between December 16, 1811 and March 15, 1812, he tallied 1,667 shocks that set the pendulums in motion but were too small to be felt, and more than two hundred shocks ranging in severity from gentle to "most tremendous."

Seismologists know that any large earthquake will be followed by a sequence of aftershocks that follow fairly well established patterns. For example, on average the largest aftershock is about one magnitude smaller than the mainshock. The rate of aftershocks also falls off with increasing time after the mainshock. Finally, the distribution of aftershock magnitudes follows a well-established pattern, with roughly ten times the number of M3 quakes as M4s, ten times the number of 4s as 5s, and so forth. As earthquakes go, aftershocks are thus relatively well behaved: we know about how many to expect; we know they will cluster in proximity to the mainshock; and we know they will diminish with time.

Every once in a while an earthquake sequence is less well behaved. We now understand that, while aftershock sequences tend to diminish with time, any individual aftershock is no different from any other earthquake of the same magnitude, and as such has a small probability of being a foreshock to a quake larger than itself. Which is to say that, just because an earthquake happens to be an aftershock, that doesn't mean it can't itself be a foreshock. Occasionally, then, an aftershock sequence can be dying down as per usual, only to eventually produce another quake as large as, or even larger than, the original mainshock. This scenario played out in the southern California desert in 1992, when the magnitude 6.1 Joshua Tree earthquake on April 23 was followed by an energetic aftershock sequence before the magnitude 7.3 Landers earthquake struck at the edge of the aftershock zone on June 28. Strictly speaking one might call Landers an aftershock of Joshua Tree, but we don't tend to do that. When a sequence produces distinct large events, we speak instead of principle mainshocks.

As unruly earthquake sequences go, the 1811–12 New Madrid sequence is epic. Around nine o'clock on the morning of January 23, 1812, another large quake struck, again generating dramatic effects along the Mississippi River valley and perceptible shaking over a very large area. For various reasons this quake remains particularly enigmatic. The distribution of documented effects suggests a location to the north of the first mainshock, but how far north remains unclear. Some researchers place the event toward the northern end of the New Madrid Seismic Zone; others, including the author, argue that it could have been as far north as southern Illinois, not far from the location of the magnitude 5.2 quake that rocked the Midwest on April 18, 2008. In all likelihood we'll never know exactly where the quake was centered. But in any case the so-called January mainshock was followed by its own aftershock sequence, generating yet another period of unrest for settlers along the Mississippi River valley.

Available accounts of the sequence suggest that the aftershock sequence following the January mainshock was considerably less energetic than the sequence of aftershocks following the December mainshock. But the worst was yet to come. Around 2:45 on the morning of February 7, 1812, the New Madrid region sprang to life again, producing what those in the area dubbed "the hard shock." This third and last of the principal mainshocks rocked the heartland once again, but it did more than that. Boatmen along the Mississippi River found themselves carried back upstream by a current that had, astonishingly, reversed its course. Elsewhere along the river boatmen found themselves tumbling down six-foot-high waterfalls where the river had formerly been well behaved. Unlike the phantom church bells of Boston, the waterfalls and reversed currents along the mighty Mississippi were real. Apart from compelling and detailed eyewitness accounts, after much sleuthing and hard work scientists now know that this quake occurred along the Reelfoot Fault, a fault that runs almost north–south, crossing the sinuous Mississippi River in three places (fig. 14.4). When the fault lurched during the earthquake the downriver side went up relative to the upriver side, creating a stair step in the sediments beneath the water. Neither the waterfalls nor the reversed current lasted long. River currents rapidly smoothed the disrupted river bottom; the water

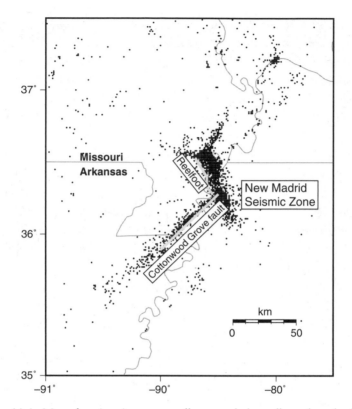

Figure 14.4. Map showing instrumentally recorded small earthquakes (black dots) and principle faults in the New Madrid Seismic Zone.

that had been sent coursing upstream eventually ran out of steam, and the river resumed its usual course. Movement on the Reelfoot Fault did leave a lasting imprint on the landscape: the upward motion on one side of the fault created a permanent dam behind which Reelfoot Lake formed (fig. 14.5).

The earliest attempts to systematically gather information about the New Madrid earthquakes, and make sense of it all, date back to the immediate aftermath of the sequence. A second eyewitness, medical doctor Daniel Drake in Cincinnati, also set out to keep a record of every felt quake and to rank them by severity. Drake further distinguished himself by not only noting that shaking was stronger along

Figure 14.5. The seeds from bald cypress trees cannot germinate in standing water. These trees, now at the edge of Reelfoot Lake, were submerged when the lake was created by fault movement during the 1811–12 New Madrid earthquake sequence. (Photograph by Susan Hough.)

the Ohio River valley than on the adjoining uplands but also correctly ascribing the difference to the difference between river valley sediments and limestone. But all of the scientific acumen in the world couldn't lead to a full understanding of the earthquakes at a time when nobody understood what an earthquake was.

By the mid-1950s seismologists generally believed that, based on their overall effects, the three principal events had magnitudes exceeding 8. The first systematic investigation of the magnitudes of the events was done by Otto Nuttli, who set out to collect and interpret historical accounts of the quakes. Nuttli was a professor of geophysics at St. Louis University, a Jesuit institution that early in the twentieth century established a top seismology department. Seismologists use the so-

called intensity scale to rank the severity of shaking from an earthquake at any given location. Unlike magnitude, which reflects the
overall energy release by an earthquake, every earthquake will generate a distribution of intensities, with highest values generally closest to
the source. By convention we refer to intensity values with Roman
numerals: III for shaking that is lightly felt, V for shaking that will start
to topple small objects off of shelves, X for catastrophic damage. De-
termining the intensity distributions for the New Madrid quakes and
comparing them to intensities generated by more recent earthquakes,
Nuttli estimated magnitudes of 7.2, 7.1, and 7.3 for the three mainshocks, respectively.

Nuttli's seminal publication appeared in the *Bulletin of the Seismological Society of America* in 1973. At a time when plans for NEHRP
were starting to gain traction, top seismologists in the central United
States were, understandably, quick to argue that resources were needed
for earthquake studies in their region. At low-magnitude 7, the 1811–
12 New Madrid quakes had been serious events, all the more so because earthquake waves travel more efficiently through the older cen-
tral/eastern North American crust than through the younger, highly
fractured crust in California. Thus church bells rang in Charleston not
only because the earthquakes were big but also because the waves they
generated traveled especially efficiently. By 1973 scientists knew that,
of all of the energy released by all of the planet's earthquakes, quakes
away from active plate boundaries account for only about 1 percent of
the total. Compared to the expected rate of damaging quakes in California, the long-term rate of damaging quakes in central/eastern
North America was known to be at least one hundred times smaller.
But if the waves from those infrequent events packed a bigger punch,
then the relative hazard might only be a factor of ten smaller.

By the late 1970s some seismologists in the Midwest had come to
believe that the 1811–12 New Madrid quakes had been even bigger
than Nuttli's results indicated. At issue was the scale that Nuttli had
used to estimate magnitude values—the body-wave magnitude scale.
By the 1970s seismologists came to realize that this magnitude scale,
like the scale originally formulated by Richter, had limitations, in par-

ticular in its ability to accurately size up the biggest quakes. By the early 1970s seismologists realized that seismic moment, a fundamentally different measure of earthquake size, was a better way to size up large earthquakes in particular. Unlike the friendly sorts of numbers that Richter's magnitude scale produces, seismic moments span a huge range. Scientists are comfortable talking about numbers like 6.3×10^{25}—63 followed by 24 zeroes—but as a rule the public is not. By the late 1970s seismologists Tom Hanks and Hiroo Kanamori formulated the moment-magnitude scale, which in effect translated seismic moment values back into equivalent Richter magnitudes. Hanks and Kanamori formulated their scale to dovetail with the earlier Richter scale for small events, but to more faithfully capture the size of big quakes.

Returning, then, to the New Madrid story, some seismologists were aware that body-wave magnitudes like Nuttli's did not generally reflect the real size of a large event. Conversions were developed to translate one magnitude estimate into the other, a translation that always increased body-wave magnitudes, often substantially. By the account of those who knew him, Nuttli resisted making this calculation. He knew what the result would be, but his gut told him that the New Madrid quakes had been big events, not what seismologists refer to as great earthquakes, magnitude 8 or larger. Eventually, and by some accounts under pressure, Nuttli made the conversion. The magnitude of the largest 1811–12 quake shot all the way up to 8.75. This estimate provided the basis for a claim that has been slow to die, that these earthquakes were significantly bigger than the great San Francisco earthquake of 1906. For all of the ruckus this result generated, it is an interesting exercise to try to find the scientific publication that describes the calculation. It was only ever published in a "gray literature" report, the kind of publication that cannot easily be found even by scientists with access to scientific libraries. Although Nuttli had in the end done the calculation, it was never published in the mainstream literature.

In early 1988, at the age of sixty-one, Otto Nuttli died of cancer. The rest of the New Madrid story played out without him.

By the time of Nuttli's death, another top young seismologist, Arch Johnston, had turned his attentions to the New Madrid sequence. Johnston was among a small group of scientists who sank their teeth into historical accounts of the earthquakes, those collected by Nuttli as well as many others collected later by other investigators. Johnston and his colleagues realized that accounts from the New Madrid area, for example those by the boatmen who had encountered waterfalls, could be used to piece together a scenario for the sequence—that is, to understand which faults had generated which event. By this time, NEHRP-funded projects had succeeded in illuminating at least some of the faults in the New Madrid Seismic Zone.

By this time, scientists also realized that all of Nuttli's calculations rested on a potentially critical weak link. Recall that Nuttli had compared the effects of the New Madrid quakes with those of other, more recent, events in the region. But all of the latter quakes were a lot smaller than the New Madrid mainshock. Thus the calculation involved an essentially blind extrapolation. To overcome this problem, Johnston set out to gather information about large earthquakes in other parts of the world that were analogous, in geologic terms, to central/eastern North America. This effort took many years; papers describing the results were eventually published in 1996. But well before the papers appeared in print, Johnston and others had a sense of the basic results, namely that the quakes had not been as big as 8.75, but had been close to or slightly bigger than magnitude 8.

News stories through the 1980s struck an ominous chord. In 1986 the 175th anniversary of the sequence sparked a number of news stories like the one published in the Frederick, Maryland, *The News*, under the headline, "Devastating Earthquake 175 Years Ago in Mississippi Valley May Occur Again." The article gave magnitudes of 8.4–8.7 for the three 1811–12 mainshocks, noting that all events were bigger than the devastating 1985 Michoacán, Mexico, quake, which had killed ten thousand people. In 1987, a month after a relatively modest magnitude 5.9 earthquake toppled chimneys and rattled nerves near Los Angeles, readers of the *Chicago Daily Herald* were again reminded, not

entirely grammatically, that "Apathy Underestimates Danger of Earthquakes" in their own backyard.

Within the scientific community many if not most seismologists
were skeptical that the principal New Madrid earthquakes had been as
large as magnitude 8, let alone 8.75. Yet the large magnitude values
continued to be touted by top experts, and found their way into the
seismic hazard maps produced by the USGS, as well as on Web sites
put together for the public. In the minds of some, hazard assessment
turned into hazard hype.

One starts to get a sense of the lay of the land in the heartland at
the time when earthquakes suddenly grabbed national headlines in the
fall of 1989. Journalists rushed to ask: "can it happen here?" The experts in the region rushed to reply. And, yet again, would-be earthquake predictors rushed out of the woodwork—one in particular who
would go down in infamy.

On November 28, 1989, the *Atchison* (Kansas) *Globe* reported that
according to climatologist Iben Browning, "quakes will rattle the New
Madrid fault in December 1990." A spattering of stories appeared in
other papers. Browning, who claimed to have predicted the Loma
Prieta earthquake, based his prediction on a projected "high cycle" of
tidal forces. Browning's career path was different from that of futurist
Gordon-Michael Scallion. Whereas Scallion had visions, Browning
had a PhD in biology from the University of Texas. Both men, however, eventually turned their attentions to forecasts of climate change,
and earthquakes. Browning believed that the earth was in the process
of cooling following a long warming phase. In writings reminiscent of
Scallion, he wrote that climate changes were associated with societal
ills such as famine, revolutions, and war.

Browning's earthquake predictions had marginally better grounding
in real science than Scallion's and were rather less vague, but vague
nonetheless. Browning identified times of high tidal stress that did
turn out to coincide with a handful of notable events, including the
Loma Prieta earthquake and the 1985 Michoacán, Mexico, earthquake.
He had also identified countless other periods of high stress that had

coincided with nothing. It was the by-now familiar story: lots of predictions; a few apparent hits; no rigorous, systematic documentation of
predictions before the fact or evaluation after the fact. In this case
Browning had given a speech to a trade association in San Francisco
on October 10, 1989; at least one individual who heard the speech
confirmed to reporters that Browning had mentioned the period October 15–17 as a likely time for a significant earthquake in the San
Francisco area. Emboldened by this apparent success, Browning began
to talk to anyone who would listen about his prediction that the next
"high cycle," on December 3, 1990, might trigger a significant earthquake in the New Madrid region.

Mainstream journalists do not generally jump to listen to would-be
earthquake predictors. A reporter from the *Columbia Daily Tribune* did
talk with Browning by phone on November 25, 1989. And the *Atchison Globe* did pick up the story, but in a short article that quoted University of Missouri–Columbia atmospheric sciences professor Ernest
Kung as saying that Browning's theories had never been proven. Most
journalists simply ignored the story; most seismologists, when asked,
continued to decry the prediction.

In early summer, David Stewart, then director of an earthquake
center at Southeast Missouri State University, weighed in, saying the
prediction was "worthy of serious and thorough consideration." A
flurry of articles appeared in mid-July 1990, after Jerome Hauer, the
chairman of the Central United States Earthquake Consortium, told
reporters: "I have taken the position that we are not going to ignore
this prediction." Hauer, who had no background in earth science, had
earned a master's degree from the Johns Hopkins School of Public
Health and by 1990 was executive director of Indiana's Emergency
Management Agency. He did tell reporters that he expected Browning
to be proven wrong, but nonetheless stopped short of dismissing the
prediction. "If we have to schedule National Guard drills during the
month of December," he said, "why not schedule them that week?"

The mainstream earthquake science community dismissed the prediction from the beginning. A lone voice of support continued to
come from David Stewart. Stewart, who earned a PhD in geophysics

from the University of Missouri at Rolla, has spent his career in what might be called the fringes of the mainstream community. In later years his interests turned to the chemistry of essential oils; he authored a book titled *The Chemistry of Essential Oils Made Simple: God's Love Manifest in Molecules.* Yet when he spoke up in 1980 with support for the Browning prediction, he spoke—at least initially—as an expert with seemingly legitimate credentials. An August *New York Times* article noted his "respect" for the prediction. Stewart's own statements later called his credibility into question, in particular when he expressed in writing his belief that psychic phenomena have been scientifically proven.

By fall of 1990 scientists in the mainstream community grew increasingly concerned about the degree of panic it was generating. In October the National Earthquake Prediction Council described Browning's method as "no more accurate than throwing darts at a calendar." Still, the story continued to garner media attention. The prediction, and response to it, began to be front-page news.

They didn't quite march elephants up main street in the town of New Madrid on December 3 but the atmosphere wouldn't have been much different if they had. Television vans arrived from near and far, some sporting enormous satellite dishes. Hap's Tavern opened early for an all-day Shake, Rattle, and Roll party. Tom's Grill served up a special Quake Burger. Another enterprising resident hawked t-shirts to visiting tourists: "It's our fault," the shirt proclaimed. An evangelical preacher drove around downtown New Madrid in a van with a loudspeaker, reassuring crowds that they had nothing to fear from the earthquake; it was Jesus Christ they should be worried about.

The prediction caused real consternation among officials and the public. Some residents left the area; others stockpiled food and supplies. Classes were canceled for the day in some area schools. In St. Louis elementary schools student attendance dropped to 50–70 percent. Law enforcement officers and rescue officials set up emergency headquarters. At the epicenter of the circus, however, most residents were amazed and amused in about equal measure. Locals joked that their greatest danger wasn't an earthquake, but of being run over by a television truck. Local businessmen, who hadn't seen as many visitors since a big Civil War reenactment a few years earlier, made the most of the opportunity. "The only things that happened was that folks around New Madrid made a boat-load of money from those 'out of towners.' Pardon my cliché, but the locals laughed all the way to the bank."

A quake might have struck New Madrid on December 3 purely by random chance, a happenstance that some experts recognized as a frightening possibility, albeit a faint one. But the day came and went with no seismic excitement. Experts in the earth science community enjoyed a measure of vindication. They had, after all, spoken out clearly and consistently to decry the prediction. Professor Seth Stein, from Northwestern University, went so far as to tape a television story in advance of December 3, explaining that the prediction was total hooey.

Some in the emergency management community pinned part of the blame for the circus on the mainstream community, in particular their failure to speak out earlier to dismiss the prediction as specious. The NEPEC evaluation had, after all, been done just six weeks prior to the time of the predicted quake, after the circus train had developed too much momentum to be halted. Some scientists, including Arch Johnston in Memphis, had spoken out earlier dismissing Browning as, essentially, a quack. Many others had preferred to stay quiet, believing it better to not dignify junk science by engaging in a public debate. Looking back some scientists expressed regret that they hadn't spoken out sooner and more strongly.

If they did not speak as quickly or as loudly as they might have, un-like the 1969 prediction fiasco in the Bay Area reputable scientists could not be accused of fanning the flames of hysteria once it got un-derway. But looking back to connect the dots, one can't help but won-der: did the mainstream community have a hand in collecting the pile of tinder-dry kindling for Browning's match to fall in?

Iben Browning made his appearance the national scene in 1989, four years after Charles Richter died. One hazards to guess that the latter man would have been among the first to describe the former man as a prime example of what he had in mind when he spoke of "fools and charlatans." The scientists who concluded that the 1811–12

New Madrid earthquakes were bigger than any quake in California in historical times are not quacks. It was a remarkable sleuthing exercise, to use the accounts of the earthquakes as well as geological observations to figure out which fault had produced which earthquake. The work to compare the earthquakes' effects with the shaking from other large quakes in similar regions worldwide was similarly an exhaustive effort. In the end the process went wrong not because the hard work was flawed, but because the ostensibly easiest part had gone awry. When researchers had assigned intensity values to the original accounts, these values were in many cases inflated. Subsequent studies had also failed to account for the fact that most of the settlers who experienced the New Madrid earthquakes were living along major river valleys, where shaking is amplified by soft sediments.

If the series of missteps that yielded the high magnitudes in the first place is understandable and forgivable, one is still left with the question of how these values were used to assess and portray hazard. From the beginning many seismologists harbored deep doubts about the very high magnitude estimates that Nuttli calculated in 1979. At a minimum, the estimates were surely highly uncertain relative to the estimated magnitude of the 1906 San Francisco earthquake, which left a clear surface fault break and had been recorded by a handful of early seismometers.

Later work has continued to highlight the extent of the uncertainties. Arch Johnston's work provided better constraint on magnitudes than earlier studies, and lowered the magnitude estimate of the largest event from 8.75 to 8.1. But it didn't take many years following the 1996 publication of Johnston's results for the scientific community to realize that these magnitude estimates were still grossly uncertain. One 2002 study looked carefully at the uncertainties and found that, for example, the magnitude of the February 7, 1812 shock could have been as low as 7.0 and as high as 8.1. While the hazard maps considered uncertainty by average calculations for several different magnitudes, the assumed range of magnitudes did not span the range 7.0 to 8.1, but rather 7.4 to 8.1. In some cases researchers came right out and said, at least in private, that people wouldn't take hazard seriously if the maps reflected the true uncertainties. They might well be right. And if people and policy makers are disinclined to worry about low-probability hazards, wishy-washy statements about the scope of the hazard are unlikely to get their attention.

Front-page news stories about earthquake predictions, those get people's attention, but not always toward the best ends. Missouri Bootheel resident Dennis Loyd and his family were worried enough about the Browning prediction to spend the night of December 2 in their camper rather than their home, only to find themselves caught in a gust of wind that sent the camper rolling. Loyd, his wife, and 16-year-old son sustained injuries severe enough to require medical attention.

The job of science is the pursuit of truth. The moment that scientists become advocates of any other cause is the moment that scientists stop being honest brokers. But who if not seismologists have the deepest appreciation for the nature of earthquake hazard? Who if not seismologists are most strongly motivated to advocate for mitigation of earthquake risk? It is a vanishingly, possibly impossibly, fine line that seismologists walk as we embrace an advocacy role: ineffective if we don't edge toward alarmist rhetoric, potentially guilty of fanning flames and fears if we do.

CHAPTER 15

Measles

And scientists now say they can calculate the probability
of an earthquake within twenty-four hours in California.
Although, they said, it's not really that useful for
predicting the first quake, but they can predict
aftershocks. Hey, don't do me any [favors]—I can predict
aftershocks, okay? You know when they happen?
Right after an earthquake.

—JAY LENO, May 18, 2008

Looking back at earthquake prediction efforts of the past one can see
how both predictions and prediction methods have gone awry. A
very small handful of credible predictions appear to have been borne
out—among them, Haicheng, the prediction of the 1989 Loma Prieta
earthquake based on the original M8 method, a small number of supposed hits by the VAN method, and the prediction of a small earthquake in upstate New York in 1973. Some respected researchers tout
other supposedly successful predictions.

The first thing that needs to be said is that if one makes enough
predictions, by scientists either individually or collectively, some will
inevitably be right. The key, again, is to establish a track record that
demonstrates better success than could be achieved by random chance,
given what we know about how often and where earthquakes tend to
occur. A single apparently successful prediction doesn't count for
much. Indeed, when Yash Aggarwal phoned his colleagues at Columbia University to report that his predicted earthquake had occurred,
one of them told the young scientist, "If you can do this three times
you will all be famous." In an article in *Time* magazine this reply was

called "enthusiastic." The trained ear detects a hint of sentiment other than unbridled enthusiasm.

But discussions in earlier chapters discuss a reason why earthquake prediction schemes will occasionally appear to be successful: the fact that earthquakes do have a tendency to cluster in time and space. In certain areas, including Greece, large earthquakes seem to have an especially strong propensity to bunch up. Thus if the VAN group, consciously or unconsciously, issues more predictions after one or more earthquakes has occurred, they might occasionally appear to predict subsequent earthquakes.

In general one will always have some success predicting that moderate-to-large earthquakes will occur after one or more moderate-to-large quakes has already struck, because almost all big earthquakes have pretty big aftershocks. As Jay Leno quipped, as prediction methods go this one leaves something to be desired. It will not ever lead to reliable, specific short-term earthquake predictions. But some predictability is better than no predictability. In recent years, a number of forecasting methods have been developed based specifically on clustering.

The simplest such method starts with a consideration of expected long-term earthquake rates and adds established aftershock statistics. For example, suppose that based on long-term rates there is a 1/100,000 chance that damaging shaking will occur along the southern San Andreas Fault in a given twenty-four-hour period. If a magnitude 6 earthquake occurs somewhere near the fault, we know how many aftershocks of various magnitudes to expect. One can calculate the probability of damaging shaking from these aftershocks, and add it to the background rate. We don't generally worry about the hazard from aftershocks of moderate mainshocks. Even for a mainshock as large as 6, the aftershocks are unlikely to pose more than a very modest additional hazard—magnitude 4 and 5 quakes are generally not the quakes we worry about, even though they can cause modest mischief.

But, as noted, increasing evidence reveals that the classic definition of an aftershock is overly restrictive; that, as the 1811–12 New Madrid sequence demonstrated, aftershocks are not always the well-behaved earthquakes the public assumes them to be. If an earthquake of a certain magnitude, say M6, occurs, the established statistics for aftershocks (one M5, ten M4s, etc.) aren't wrong per se, but rather incomplete. In particular, once that M6 quake happens, there is increasing evidence that we should think of it as a "parent" quake that can beget "daughter" quakes of any magnitude. On average the biggest daughter will be M5. But some parents will produce fewer big daughters than average, and some might be especially prolific. A daughter quake could be the same size as a parent quake; a daughter quake could also be bigger.

It turns out that if one simply takes the established probabilities for aftershocks and extends them to include daughters that are bigger than parents, the statistics match what we see. So, then, if we add aftershock probabilities to hazard based on the background rate, we can extend the probabilities to include the possibility of an especially strapping daughter. The odds of this are low, but they still represent a significant increase to the probabilities based on background rates.

Various methods have been developed based on this idea. Fundamentally the approach is straightforward, drawing on established statistics from past earthquake sequences. But one can develop methods in ways that are rather less straightforward. Seismologist John Rundle and his colleagues have spent years developing what they call the Pattern Informatics, or PI, method. Rather than relying on simple characterizations of aftershock or daughter quake statistics, the PI method defines a more complex measure of earthquake activity in a given region. If activity goes up or down significantly in a certain area, according to the method future quakes are more likely to occur in that region. The method is mathematically complex, but in the end spits out maps with blotchy measles spots that indicate regions of increased probability. When one looks closely at the measles, they overwhelmingly fall close to areas where small or moderate earthquakes have happened recently. That is, while in theory the measles spots could fall in areas that have been conspicuously quiet relative to long-term rates, in practice they rarely do.

Rundle and his colleagues have not been shy about touting their success in the media, for example following the 2004 Parkfield earthquake as well as the M6.5 San Simeon quake, which struck a different part of central California in 2003. Days after the Parkfield quake Rundle's colleague Donald Turcotte told a reporter, "The fact that these earthquakes have occurred in areas that were forecast five years ago has to be considered quite remarkable." The researchers touted the method as a way to identify regions at heightened risk of large quakes, and therefore regions where retrofitting efforts might be stepped up.

But the distinction between forecasting and prediction starts to blur. Following the 2004 Parkfield earthquake, Turcotte called into a show on radio station KQED in San Francisco. Listeners, he said, might be interested to know that he and his colleagues had successfully predicted the recent quake as well as other significant quakes in California. The difference between prediction and short-term forecasting might sound like a distinction in semantics, but of course it is much more than that. Whatever precise terms one uses to define a meaningful short-term earthquake prediction, the public has at least a general understanding of what earthquake prediction is about. In short, prediction is saying when and where an earthquake will strike, and how big it will be. Producing a map that shows that future earthquakes are more likely to strike near one of a large number of blotchy measles spots is not what the public understands to be earthquake prediction.

Whether the PI method represents any improvement over the much simpler short-term forecast methods that consider well-established statistics of parents and daughters is difficult to say. Whereas the simpler methods based on earthquake statistics produce statistical forcasts that can be tested, the measles map approach has not lent itself to rigorous evaluation. Considering the statistical performance of the method, another team of researchers, John Ebel and Alan Kafka, concluded that the PI method performed worse than a much simpler "cellular seismology" approach to forecast earthquakes based on short-term clustering.

But if none of these methods can hope to lead to anything more than fairly weak statements that large quakes are more likely, one might ask: what good are they? Stripping all of the complications aside, established foreshock statistics in California tell us that an earthquake of any magnitude has a 5 percent chance of being followed by something bigger within three days. No matter how one looks at parents and daughters, in the end one arrives back at this very simple, and not enormously useful, statistic. So far as we understand, of every twenty magnitude 6 quakes that occurs close to the southern San Andreas Fault (say), we expect only one of these to be followed by something bigger. And even then the odds are that the bigger quake will not be that much bigger.

So what does this approach get us? In practical terms even a weak probability increase does have some utility. A one in twenty chance might be low, but it is high enough to justify some action. For example one wouldn't evacuate Palm Springs based on a one in twenty chance of a large earthquake but one might reasonably move fire engines outside of fire houses for a few days, to make sure they don't get trapped if a quake occurs and the fire house is damaged. Seismologists and emergency managers can be on alert. And so forth.

These short-term forecast methods are also important for scientists' continuing quest to understand earthquake predictability, in particular to evaluate proposed earthquake prediction methods. Consider the VAN method, which as discussed is based on the idea that the earth generates characteristic electrical signals prior to large earthquakes. As a prediction method, most seismologists consider VAN to have been resoundingly debunked. The remaining question is, if one or more specific predictions appear to be successful, is it because the earth does generate these signals, or because the method manages to benefit from the tendency of quakes to cluster. In short, are the fundamental scientific underpinnings of the method valid, or are they complete hogwash? To answer this question, for VAN or AMR or M8, one needs a basis of comparison. The parent-daughter methods provide that basis. They tell us how successful a prediction method can be on average given only established earthquake rate statistics.

If a day comes when a method performs statistically better than a parent-daughter method we will have some basis for confidence in not only the method but also the physics on which it is based. In recent years scientists have begun to make such comparisons. They reveal that, occasional bold statements and press releases notwithstanding, we are not there yet.

CHAPTER 16

We All Have Our Faults

> All parts of the globe have been at some time visited by
> earthquakes and volcanic eruptions; and those portions of
> the sphere where the interior fires have most to feed
> upon, and where they reach their fiery tongues nighest
> the surface, sometimes break forth in volcanic eruptions.
> —R. Guy McClellan, *The Golden State*, 1876

In a sometimes flamboyant 1876 history of the state of California, Guy McClellan describes the geological paroxysms that had struck other parts of the country and the world, observing that "compared to the earthquakes of other times and countries, California's earthquakes are but gentle observations."

This book focuses on California because, McClellan's pronouncement notwithstanding, the state has long been a playground for earthquake science and earthquake prediction research. Within the United States, Alaska has more earthquakes and bigger earthquakes, but with its remote location and sparse population cannot match California for either quality or quantity of data or societal impetus to understand earthquake hazard. Within the contiguous United States the Pacific Northwest is also now known to have bigger quakes than the Big Ones we expect in California, but the major plate boundary fault is offshore and underwater. The region has also experienced a lower rate of moderate and large earthquakes in historical times than California. Along with a handful of places around the world, including Japan, Turkey, and New Zealand, California is a natural laboratory for earthquake science. The San Andreas is the major plate boundary fault, but the entire state can fairly be considered a plate boundary zone.

The plate boundary extends north of California, of course. We now know that the Cascadia subduction zone produced the largest quake witnessed in the contiguous United States in historical times, a massive magnitude 9 shock on the night of January 26, 1700. (McClellan might not have been so wrong after all.) The native peoples who witnessed this portentous event did not keep written records; instead, they relied on an oral storytelling tradition to pass the information from one generation to the next. The tsunami generated by the 1700 quake also reached the shores of Japan, where residents of the coast kept tide gauge records that ultimately provided the key for identifying the precise date of the earthquake (fig. 16.1).

In recent years geologists have undertaken extensive investigations of buried tsunami deposits along the Pacific Northwest coast. These studies reveal that great earthquakes like the 1700 quake strike about every five hundred years on average. As scientists work to improve forecasts, they once again run up against the problem of irregular clocks. In recent years there has been optimism that occasional slow slip events might someday hold the key to refined short-term hazard estimates, if not earthquake prediction. Along subduction zones, these slow slip events involve movement on deep patches of the main subduction zone fault, below the depths at which earthquakes occur, that occurs on a time scale of weeks. For example it is possible that the odds of a megaquake are much higher during the times when a slow slip event is underway. Whether the optimism will be borne out this time remains to be seen.

When one considers earthquake hazard along other active plate boundaries, the issues and challenges are largely similar to those in California. These include a number of the plate boundaries we worry about because of the danger they pose to population centers: the Himalayan arc, the front line of the collision between the Indian subcontinent and Eurasia; the subduction zones offshore of Mexico, Japan, Central and South America; the North Anatolian fault in Turkey. In all of these places we have some understanding of the irregularity of local clocks, some understanding of expected long-term earthquake rates.

It remains an enormous challenge, even in the active plate bound-

Figure 16.1. The last great earthquake on the Cascadia earthquake caused large tracts of land to sink. Tidewater rushed in, inundating and killing large stands of trees. Over time sediment accumulated in these areas, creating marshes and, later, coastal meadows. Today, so-called ghost forests remain, standing in silent testimony to powerful forces that reshaped the landscape over 300 years ago. (USGS photograph, courtesy of Brian Atwater.)

ary zones for which we have the most extensive information, to move from long-term forecasting to meaningful short-term forecasting, let alone prediction. That challenge pales in comparison with the challenges scientists face in trying to understand and predict earthquakes in the parts of the planet that are not active plate boundary zones—which is to say, most parts.

In regions like Australia and the central United States, big earthquakes happen much less frequently than in Japan and California. This is, of course, mostly good news for those who live in Australia and the central United States. But it's bad news for the cause of understanding earthquake hazard; we have far less data to work with because we've seen far fewer earthquakes during the historical record. What's more, increasing evidence suggests that earthquakes strike by especially irregular clockwork in these places, with especially pronounced clustering in time. For example in New Madrid we have good geological evidence that at least three sequences of large quakes have struck within the past one thousand or so years: around AD 900, AD 1450, and in 1811–12. But scientists are able to look at the deeper, older strata in the region, and these layers tell us that large earthquakes have not been happening in this region for millions or even tens of thousands of years. At some point in the past ten thousand years the New Madrid zone turned itself on; at some point presumably it will turn itself off.

To the north of New Madrid, along the Wabash Valley zone between Illinois and Indiana, geological evidence tells us that several large earthquakes occurred several thousand years ago. Did this zone switch on and off before New Madrid came to life? One plausible theory says that earlier earthquakes in the Wabash Valley and more recent quakes in New Madrid are due to the slow rebound of the earth's crust following the melting of the Ice Age ice sheets that covered northern North America. But we are left with some pretty big questions. For example, has the activity run its course at New Madrid? When scientists use GPS measurements to look at the region they find no evidence for increasing stress. Maybe the switch has already been switched.

And, are we so sure the Wabash Valley has turned itself off? Through the twentieth and early twenty-first century the Wabash Valley has had

Figure 16.2a. Damage in Charleston, South Carolina, from 1886 Charleston earthquake. (USGS photograph.)

far more moderate quakes, with magnitudes close to 5, than the New Madrid zone.

Scientists are left with similar questions in other parts of the United States. For example we know compelling geological evidence that pretty big quakes, around magnitude 7, have struck the coast of South Carolina near Charleston, on average about every four hundred to five hundred years (figs. 16.2a and 16.2b). Is the next big quake unlikely because the last big one occurred barely over one hundred years ago? And what's so special about Charleston, anyway? In geological terms one is hard-pressed to distinguish the region from the rest of the Atlantic Seaboard. Could the next magnitude 7ish quake be in an area that is now stone-cold quiet? In the Chesapeake region, say? Or near New York City? Or Boston? Damaging quakes have struck near Boston during historical times, including the Cape Ann earthquake of

Figure 16.2b. Damage to St. Phillips Church in Charleston, South Carolina, from the 1886 Charleston earthquake. The bell in this same church was set ringing by the largest of the New Madrid earthquakes in 1812. (USGS photograph.)

November 18, 1755, which was strong enough to throw down many chimneys.

One of the biggest quakes in historical times in eastern North America struck back in 1663, an event on February 5 along the St. Lawrence Seaway near the Charlevoix region of Quebec. Other moderate quakes have struck in or near Charlevoix more recently, including the magnitude 5.9 Saguenay earthquake in 1988. But could the next big quake strike to the south, closer to Quebec City?

Are there other zones out there, lurking quietly throughout historic times, that will turn on at some point and become the locus of activity for the next few thousand years? Scientists are starting to think that the answer is yes; that in places like the central United States, earthquakes happen in flurries that last perhaps a few thousand years. Other parts of the world are in the same boat, including northern China, peninsular India, Australia, central Canada. In these regions large earthquakes are unlikely, but those big quakes that do strike are especially likely to surprise us—and to cause catastrophic damage in regions that are unprepared.

Complicated plate boundary zones also pose a challenge. Of the world's active plate boundary zones, the collision between the Indian subcontinent and Eurasia is arguably the most complicated—or at least, the most complicated on a grand scale. India began to collide with Eurasia about 40 million years ago. Since that time the two land masses have been stuck together, but forces deep in the earth continue to push India northward. Thus stress continues to build along the Himalaya arc; stress that is released in large earthquakes such as the Kashmir quake of 2005 and the Bhuj, India, earthquake of 2001 (fig. 16.3), and, we now know, occasional much larger quakes.

The ongoing northward push of India also makes itself felt over a much bigger area than the extent of the Himalayan arc. The Tibetan Plateau has been pushed upward, but to some extent it is also now being pushed out of the way. Unfortunately there is no place for it to go except into the rest of China. Thus there are several large, active fault zones throughout western China. One of these, the Longmenshan fault zone, was responsible for the M7.9 quake that struck the Sichuan Province on May 12, 2008 (fig. 16.4). We know that the rate of large earthquakes in places like southwest China will be higher than places like northwest China that are not close to active plate boundar-

Figure 16.3. *Top, facing page.* Damage from the 2001 Bhuj, India, earthquake reflects not only the severity of shaking but also the vulnerability of local construction. The relatively well-built temple in the background survived unscathed in a region where many other structures were completely destroyed. (Photograph by Roger Bilham, used with permission.)

Figure 16.4. *Bottom, facing page.* Damage from the 2008 Sichuan, China, earthquake. (Photo courtesy of the Earthquake Engineering Research Institute.)

ies. But we also know that big quakes will be distributed over a much bigger area than along simpler plate boundary zones like Japan and California.

As far as earthquake prediction research goes, clearly some of the methods discussed in early chapters can and have been applied in any part of the world. If, for example, faults did produce anomalous electromagnetic signals prior to large earthquakes, one could look for those signals anywhere and everywhere. As scientists have worked to develop earthquake prediction methods they have focused on places like California, Japan, and Greece in part because of societal impetus, but in part because we have the best hope of making progress where we have the most, and the best, data. If we want to try to look for earthquake patterns that might herald an impending large earthquake, we want to work with the best earthquake catalogs we can find. If we try to look for patterns in areas with fewer quakes and/or fewer data, the challenges only go up.

All of which is to say that, if earthquake prediction research has not been encouraging in places like California, it is more discouraging still for places like New Madrid. The clocks are much slower, much more irregular, and much less well understood. Back in the early 1980s seismologists dared to predict that the next Parkfield earthquake would occur within four years of 1988. That quake did eventually happen, but it was twelve long years late. Based on the best geological studies in the New Madrid region we might expect the next big sequence to be sometime around 2250. If it's as proportionally late, relative to the averages, as Parkfield, it might not happen until 2400 or thereabouts. If the clock is more fundamentally irregular it might not happen for a million years.

Ironically, arguably the impetus to pursue earthquake prediction research is much stronger in places like the central United States, central India, and northern China, where we know much less about long-term rates, than in places like California, where we're pretty good at mapping out where large quakes will strike. Californians might be keen to know when the next big one is going to strike, but quakes strike often enough to be recognized as a real and present danger. In most of the world, where earthquakes are not common, we currently have no way to know which of innumerous relatively inactive faults might produce the next damaging quake. If the earth does produce precursory signals that tell us a big earthquake is on the way, those signals would be especially useful for averting disaster in the quakes that stand to surprise us.

CHAPTER 17

The Bad One

Everyone agrees that Los Angeles and San Bernardino
should treat this like a final warning. It's like when you
clean up camp. It's time to make that last pass through
our cities, homes, and lives and act as if the damn
thing will happen tomorrow.
—Allan Lindh, 1992

So what about California, anyway?

It has been about a century since geologists first followed the trace
of the San Andreas Fault in southern California. It has been about a
half-century since scientists worked out the theory of plate tectonics,
which tells us that the San Andreas is the principle plate boundary
fault in California. It has been about a quarter-century since geologists
began digging trenches across the fault to piece together chronologies
of past earthquakes on the fault.

Throughout this time, since the earliest days of earthquake exploration in southern California, earthquake predictions have emerged from
the pseudo-science community. But science has also cried wolf. Bailey
Willis in the 1920s. A big cast of characters, including top leaders in
the field, in the 1970s. Keilis-Borok in recent years. Apart from the
handful of notorious, specific predictions, many more earth scientists
have spoken about the southern San Andreas Fault in dire terms. According to some scientists the fault has been "ten-months pregnant"
for years now, going on decades. Scientists were worried in 1986, after
the moderate, M5.9 North Palm Springs broke a small segment of the
San Andreas in the desert. We were worried in 1992, when it appeared
that the Landers–Big Bear sequence had unclamped a section of the
central San Andreas Fault near San Bernardino. Some scientists were

worried when GPS instruments revealed unusual signals in early 2005. Some of these same scientists expressed concern over the apparent increase in the rate of magnitude 3–5 earthquakes in southern California in 2008 and early 2009.

The field of earthquake science has made impressive, collective strides forward in recent decades. We understand how often earthquakes have to occur on the fault—the San Andreas proper as well as subsidiary faults like the San Jacinto and Hayward—to keep up with the relative motion of the North American and Pacific plates. We can pinpoint the timing of previous large quakes on these faults, in some cases going back a millennium or more. We know the date of the most recent large quake on many faults, if not from the historical record then from carbon-14 dating. To recap, the northern San Andreas last broke in 1906. The central stretch of the fault broke in 1857. The southern stretch, from roughly San Bernardino to the Salton Sea, last broke some time around 1690, an earthquake that was not witnessed by people who kept written records. In northern California the Hayward fault, which bisects the heart of the now densely populated East Bay area, last broke in 1868.

Starting in 2004 a group of scientists undertook a major effort to reassess the probabilities of earthquakes in California, the latest in a series of working group efforts that date back to the late 1980s. The project was funded by the California Earthquake Authority, an agency that has a mandate to provide earthquake insurance with rates set according to "best available science." Led by USGS seismologist Ned Field, the project drew from the expertise of dozens of leading experts, incorporating results from hundreds of recent studies. When one puts together a so-called earthquake rupture forecast, one can develop what seismologists call a time-independent model—which is to say, an assessment of probabilities based only on the average long-term rate of earthquakes on different faults. In such a model, then, the expected rate of quakes on the northern San Andreas would be determined only by the average rate as revealed by geological investigations—the fact that the last big quake struck in 1906 would not enter into the calculation. Field's group put together such a model.

The group also pressed one step further, assessing statewide probabilities taking into account known dates of last big earthquakes. This calculation revealed the faults that, according to the collective wisdom of the field, are starting to look especially ripe. Still one hesitates to return to the "*o* word." Overdue is an unfortunate word in this context not because it is wrong, but because it has the wrong connotations. When a baby is overdue, it will arrive in at most a couple of weeks. When a bill is overdue, it will get paid in at most a month or there will be consequences. As earlier chapters have discussed, when an earthquake on a certain fault is overdue according to the law of averages, it might still be fifty or even one hundred years away.

If science can say to a near certainty that a large quake will strike on a given fault within the next fifty years, this is a tremendously useful forecast. In fact some scientists believe it tells us everything we need to know as a society to mitigate risk: it tells us that every house, every school, every hospital, every freeway overpass, every mini-mart, every structure, and every piece of infrastructure needs to be designed and built to withstand earthquake shaking. The traditional approach to building codes has not been to ensure the continued functionality of a structure after a strong earthquake but rather to provide for life safety. That is, a structure built to code might not come through the next Big One entirely unscathed; the point of the codes is to make sure its occupants do. As the cost of construction and reconstruction, and the value of urban building stocks, have skyrocketed, some experts have questioned the traditional approach, wondering if, as a society, we can afford a big quake that comes with a $100 billion, or a $500 billion, price tag.

If earthquake science could perfect the art of forecasts on a fifty-year scale, we would know what structures and infrastructure would be up against. For the purposes of building a resilient society, earthquake prediction is largely beside the point. Whether the next Big One strikes next Tuesday at 4:00 p.m. or fifty years from now, the houses we live in, the buildings we work in, the freeways we drive on—all of these will be safe when the earth starts to shake, or they won't be.

As sentient individuals who don't appreciate having the ground beneath our feet becoming unglued, we worry about next Tuesday; we don't, as a rule, worry much about what will happen fifty years from now. But we—human beings as well as scientists—really would like a heads up if the Big One is going to strike next Tuesday at 4:00 p.m. We want it so badly that even trained scientists are sometimes not immune to false hopes.

Some respected scientists argue that earthquake prediction will never be possible. This school of thought derives a measure of support from various observational and theoretical studies. For example, some scientists who view earthquakes from a perspective of statistical physics conclude that earthquakes do pop off like popcorn kernels, most destined to remain small, a very few destined to explode beyond the realm of proper popcorn etiquette. Accordingly, the problem of earthquake prediction reduces to the problem of predicting which few of the many regular popcorn kernels will explode into superkernels. But if a kernel doesn't know how big it's going to get until it starts to explode, prediction will be hopeless.

Another interesting model, proposed by Nadia Lapusta and her colleagues, suggests that earthquakes will nucleate in especially weak patches along faults, what Lapusta calls defect zones. Within this framework, a large earthquake occurs when a small earthquake is able to push its way into a neighboring, stronger part of the fault. Once a rupture gets going the earthquake serves to weaken the normally strong fault, and thus tends to keep going. If this is how earthquakes work, the key for prediction would again be figuring out which small quake among many small quakes will succeed in busting out.

Many respected scientists stop short of unbridled pessimism. If we don't know exactly what goes on in the earth before stress is released in a large earthquake, who is to say that the process isn't accompanied by some sort of change we might be able to detect in advance. Maybe large quakes are set off by the kind of slow slip events that scientists have observed along subduction zones. Maybe small and moderate quakes in a region do follow characteristic patterns as a major fault nears the breaking point. Maybe mineral alterations deep in the earth's crust do release fluids that percolate through rock and eventually trigger big quakes. Maybe rocks do start to crack as stresses on a fault reach a breaking point.

And maybe the ideas touted by amateur earthquake predictors will someday be shown to have a measure of validity. Maybe tidal stresses or atmospheric pressure has something to do with earthquake nucleation. Maybe gasses, fluids, or heat are released from the ground in advance of a large quake. Maybe the earthquake prediction puzzle will someday be solved by an approach that isn't a glimmer in anybody's mind at present.

Where earthquake prediction is concerned, there is no shortage of maybes. Scientists continue to pursue many if not most of the ideas discussed in this book to formulate prediction and short-term forecasting methods, and to tackle the enormous challenge of evaluating the success of those methods.

Earth scientists also continue to sink their teeth into active faults, to identify the geological evidence of past earthquakes and to better estimate how often big quakes strike on key faults. This kind of work more than any other tells us what we are up against: whether big quakes on the San Andreas Fault strike on average every 150, or every three hundred years; whether those big quakes are always about the same size, or whether they are mostly relatively small (magnitude 7ish) and only occasionally especially large (magnitude 8ish).

When paleoseismology investigations were first undertaken in the closing decades of the twentieth century, they suggested that large quakes on the northern, central, and southern San Andreas occur on average about once every two hundred years. More recently, as geologists have dug deeper into these faults and used more precise dating techniques, these numbers have been dropping. A key point here is that we do know how fast the plates move, and this provides a fundamental constraint on the total amount of energy that earthquakes will release. If the earthquakes are relatively more frequent, then they must be relatively smaller—kind of a good news, bad news situation.

But as scientists have focused their attentions on the southernmost San Andreas Fault system, the bad news seems to pile up. From about San Bernardino southward, various lines of evidence suggest that it probably isn't right to think of the San Andreas as a simple plate boundary. Rather the San Andreas and San Jacinto faults appear to be something close to equal partners. We might be witnessing a snapshot of an evolutionary process: the relatively younger San Jacinto Fault might be in the process of taking over as the major plate boundary fault, a process that will continue for hundreds of thousands, if not millions, of years.

But regardless of how earthquakes will divvy themselves up between the southern San Andreas and the San Jacinto faults, one thing is clear: from the start of the historic record (circa 1769) to the time of this writing, neither fault has produced a great earthquake. Detailed investigations by geologist Tom Rockwell and his colleagues tell us that the most recent big quake—big enough to leave a surface scar—on the San Jacinto fault was in the mid- to late-18th century.

Sometimes it seems like scientists don't agree on anything, but for the southern San Andreas system the results appear to be beyond dispute: it has been a very long time since the last big quake. Eventually the logjam has to break. It could break in stages. According to geological evidence and the law of earthquake magnitude averages it is more likely to break next in a relatively small event, closer to magnitude 7 than 8. An earthquake like this would be bad news for desert communities closest to the fault, for example Palm Springs, Indio, and Mexicali and Calexico to the south. Such a quake could also be more broadly disruptive, for example to the railway lines that parallel the southernmost San Andreas Fault, the primary freight conduit for goods from the port of Long Beach to points inland.

But a lot of strain has built up in the 250 to 300 years since the last Big One in this part of the world, and earth scientists worry that the next Big One could be really big. The worst-case scenario would be the wall-to-wall rupture of the southern San Andreas, unzipping the entire southern San Andreas from Parkfield to the Salton Sea. Short of the worst case are several nearly worst cases, for example a quake that makes it two-thirds of the way from wall to wall—an earthquake of magnitude 7.8, or thereabouts.

A bigger Big One worries us for several reasons. First, we know that shaking is most severe close to the earthquake source, so the longer the earthquake, the more real estate directly in harm's way. The Southland's desert communities, including Palmdale and Lancaster to the north of Los Angeles, have grown a lot in recent decades.

Second, there is the matter of lifelines. At Cajon Pass alone one finds a spider web of key lifelines: power lines, rail lines, high-pressure gas pipelines, and more. If the two sides of the San Andreas are suddenly yanked sideways relative to each other by five or ten meters, all of these lifelines will be yanked apart, with all-too-predictable consequences. Ruptured gas lines will explode, power lines will fall, rail lines will be twisted and broken; the key I-15 highway artery, built atop a large earthen abutment where it crosses the fault, will be left in shambles. Fires will start; how far they spread is impossible to know, but clearly if the Big One happens at the wrong time, the firestorms could by themselves be catastrophic. Even away from the major lifeline crossings, every power line, every transformer, every propane tank is a potential ignition point. And every water pipeline, not to mention every reservoir, will be a point of vulnerability. During the Santa Ana weather conditions, southern California can be hot, windy, and tinder dry; it doesn't take an earthquake to send the Southland up in smoke.

Pipelines can, at considerable expense, be built to withstand shifts across faults. When the designers of the Trans-Alaska Pipeline realized that their pipe had to cross the active Denali Fault, they engineered an ingenious solution that allowed sections of the pipe to move on skids (fig. 17.1). The investment paid off when a magnitude 7.9 quake struck on the Denali Fault on November 3, 2002. The pipeline sustained some damage but did not rupture, averting what would have otherwise been a massive economic and environmental disaster. It was and is nothing short of a poster child for the cause of earthquake risk mitigation, the up-front cost that pays for itself many times over. But the Alaska pipeline was a high-stakes game: big investment, big pay-off. For run-of-the-mill pipelines, and pipeline owners, this degree of mitigation is viewed as prohibitively costly. The strategy is thus to let the pipelines (also the canals) break and then fix them as quickly as possible (fig. 17.2) Meanwhile earthquake experts cast a wary eye on the web of critical arteries that connect the greater Los Angeles metropolitan area to the rest of the country.

Returning to the potential impact of an especially big Big One, the

Figure 17.1. The Trans-Alaska Pipeline, built over a series of skids where the pipeline crosses the Denali Fault. When the magnitude 7.9 earthquake struck on this fault in 2002, the fault moved seven meters (twenty-three feet) within the fault zone that geologists had identified, but the pipeline did not break. (USGS photograph.)

experts have also begun to consider, and worry about, the potential impact on the urban population centers, not only San Bernardino but also Los Angeles and its environs. We know that a big quake, even if not directly in the LA area, will send waves into the giant bowls of sediments that underlie the flatlands of Los Angeles, including the out-lying valleys and basins. We know those waves will get trapped, slowed down, and amplified in amplitude. People who have experienced earthquakes sometimes talk about "rollers" versus "shakers." The dis-tinction usually relates less to earthquake size than to the distance of an observer to a quake. In particular, if one is living on a giant bowl of

Figure 17.2. Gas pipeline emerges from the ground where it crosses the San Andreas Fault near the town of Cholame, California. (Photograph by Susan Hough.)

sediments, as many of us do, one experiences mostly the sloshing of waves in the geological bathtub, not the more jerky motion that dies out as it travels through the earth's crust.

At magnitude 7.3, the 1992 Landers earthquake was big enough to cause serious sloshing in basins and valleys throughout the greater Los Angeles metropolitan area. Even though the quake was centered over 160 kilometers (one hundred miles) away, the still powerful waves provided a heart-pounding 4:57 a.m. wake-up call for millions. Houses rocked back and forth as if shaken by a giant, unseen hand. Power lines arced, transformers exploded, car alarms blared. The strong shaking continued for what seemed like minutes. Yet, for all of the drama, and all of the power, and all of the noise, when the dust settled residents in and around Los Angeles looked around to find a world pretty much as it was when they went to bed the night before. The waves we felt that morning were akin to the lowest tones produced by the deepest bass instruments. They might reach straight into your chest, but, at least in that case, they weren't the kinds of tones that damage houses and small buildings, and they weren't big enough to damage larger structures.

If a similar-sized quake strikes the southernmost San Andreas near Indio, the effect in Los Angeles probably won't be too different from what happened in 1992. But if a quake is significantly bigger, then it will involve the section of fault that is closer to Los Angeles, and it will start to generate a lot more of the low tones that can set tall buildings swaying. What happens then is a matter of some debate. When the 1857 earthquake sent booming waves into the LA Basin, "The ground immediately around us seemed to shake violently like a cradle rocking," wrote one eyewitness. "Only rarely do earthquakes last so long and have such strange motions." Houses and other small structures fared relatively well: two houses were reportedly knocked down in San Fernando, north of Los Angeles, but the Mission buildings escaped unscathed. No tall buildings were around to sway in response to the long and "strange motions."

The powerful quake along the west coast of Mexico in 1985 generated similar shaking in the valley beneath Mexico City, a distance of over two hundred miles from where the quake was centered. In that case the valley acted like a tuning fork, ringing at tones that were especially efficient at shaking twenty-story buildings. Although damage in the city was light overall, over four hundred buildings collapsed. Estimates of the death toll vary from 10,000 to as high as 35,000. Some engineers believe that this scene will not be replayed when a Big One strikes close to Los Angeles, pointing to California's more stringent building codes and better building practices.

Other experts aren't so sure. The California Seismic Safety Commission estimates that the state has some forty thousand nonductile concrete buildings. Constructed as stores, schools, parking structures, and office buildings between the 1930s and the early 1970s, these buildings are typically two to twenty stories tall, with the weight of the building borne by unreinforced or poorly reinforced pillars. In the words of Caltech engineering seismology professor Thomas Heaton, "It's well recognized within the earthquake professional community that many California non-ductile concrete buildings are at unacceptable risk of collapse in moderately strong shaking."

Some experts worry about more modern buildings, built to much more stringent codes. The 1994 magnitude 6.7 Northridge, California, earthquake revealed a previously unsuspected vulnerability of steel-frame buildings. Although no such buildings collapsed, inspections revealed significant damage to the welds between horizontal beams and vertical columns. Building codes were revised, and made more stringent, in 1997. But as usual one was left with the prohibitive expense of retrofitting existing structures. The costs associated with retrofitting vary enormously based on the size and construction of the building, but as an example, in 2002 Los Angeles County estimated it would cost $156 million to bring just four of its hospitals up to code under a state bill passed in the aftermath of the Northridge earthquake.

Recent studies have considered the earthquake safety of California's most modern steel-frame and reinforced concrete structures, those built to the stringent current code. To test the performance of buildings in earthquakes scientists can do one of three things: they can subject scale models to actual shaking on a so-called shake table; they can make simple calculations based on the overall size, "stiffness," and so forth, of a structure; they can develop a sophisticated computer simulation of a building and subject it to simulated shaking. Shake-table tests are useful but only to a point. Apart from the question of prohibitive costs, one obviously can't build a twenty-story building on a shake table, and a scaled-down model would respond differently to earthquake shaking than the life-size structure. Meanwhile, simple calculations tend to fail because they don't capture the complexities of building response to shaking. For example, buildings often fare better than one might predict from a simple calculation because, essentially, when a real building creaks and cracks in real shaking, every creak and every crack represents a dissipation of energy. Those creaks and cracks add up, draining energy that could otherwise set the building swaying. In fact, in some earthquake-prone parts of the world, for example Turkey

Figure 17.3. Traditional buildings in Srinigar, Kashmir, incorporate wood elements and masonry infill. The structure in the middle of the photograph remains standing, and in use, even though it is clearly compromised. (Photograph by Susan Hough.)

and the Kashmir Valley, people have long recognized creaks and cracks as an effective defense against earthquake damage. Traditional architecture in these regions incorporates a patchwork quilt of wood elements and masonry infill, producing buildings that are able to dissipate shaking energy in a million little internal shifts and shimmies (fig. 17.3). (Traditional architecture elsewhere in the world, for example Incan construction in South America, also incorporates elements that, one suspects by design, serve to improve resilience to shaking [fig. 17.4].)

The most sophisticated technique for predicting building response involves sophisticated computer simulations and high-powered computers that can capture more of the complexities of building response. Although still very much a developing research avenue, some of the

Figure 17.4. Reconstructed Incan structure at Machu Picchu in Peru incorporates timbers that, by design or otherwise, serve to brace the stone walls. (Photograph by Susan Hough.)

Figure 17.5. Damage to the I-14/I-5 interchange from the 1994 Northridge, California earthquake. (USGS photograph.)

first such simulations have produced unsettling results, namely that even modern mid-rise buildings in Los Angeles could collapse if a Bad One strikes on the San Andreas.

This is scary stuff. By now we have all seen horrific images of earthquake damage beamed into our living rooms in living color. Flattened villages in India and Pakistan; collapsed low- and mid-rise buildings in Mexico City and Turkey; schools reduced to rubble in Sichuan. We've seen images from the United States as well: the Nimitz Freeway pancaked, freeways in Los Angeles ripped apart (fig. 17.5), apartment buildings collapsed. But Americans don't imagine that shiny, modern tall buildings will ever fall down. Nor do we imagine that dozens or even hundreds of modern mid-rise buildings will collapse.

It is important to note that the jury is still out on what will happen when a big Big One strikes on the San Andreas. As sophisticated as they are, the best computer models not only can't capture the enormous complexity of real building response, by necessity they can only use predictions of what the shaking will be. And, again, scientists don't know just how big the next Big One will be. Our computer simulations reveal that, among other uncertainties, shaking in the Los Angeles area will depend significantly on where the earthquake starts.

And so here we sit, with reason for concern but with grossly imperfect knowledge. What do we say? What can we say? There's a bogeyman out there; we know it's scary and dangerous but how scary and dangerous, we really don't know. And exactly when it will make an appearance, we don't know that either. We believe it will be sooner rather than later, but "sooner" in a geological sense can still be "later" in the sense that humans care about. Once again the line between responsible communication of concern and overly alarmist rhetoric is fine to the point of invisibility. We are, for all intents and purposes, where Bailey Willis was in the 1920s: if we speak with measured concern then people don't listen, if we overstep with talk of dire predictions then we might get people's attention, but we can find ourselves discredited.

From the time seismologists moved to southern California and set up shop, they were impelled by a sense that a big quake was likely to happen. The sense of urgency has waxed and waned over the decades, but has come to the fore on at least a half-dozen different occasions.

The earth continues to give up its secrets slowly. To understand how frequently earthquakes have occurred on a given fault requires painstaking investigation of the geological signatures of past earthquakes— where they can be found at all. As far as earthquake prediction goes, scientists continue to explore some lines of evidence that appear to have some promise. History suggests that these promising avenues might eventually tell us something useful about earthquake processes, but aren't likely to live up to expectations.

History further suggests that, whether science is right or wrong about the urgency of the earthquake problem, debates will never be settled, and points will never be driven home among the public and policy makers, until the earth settles the issue for us. Will tall buildings collapse in Los Angeles when the next Big One strikes on the San Andreas? We won't know, perhaps we can never know, until the earth does the experiment for us.

What day this will be, the truth is, we don't know that either. "It could be 20 years from now, it could be tomorrow"—such is the mantra repeated by earthquake scientists in the 1920s, the 1970s, and today. Our gains in scientific knowledge leave us with a much better understanding of how often large earthquakes strike on the San Andreas, the San Jacinto, and the Hayward faults. But they also leave us with a much better understanding of just how capricious earthquakes can be, of how much we still don't know.

After the 1992 Landers earthquake rumbled through the southern California desert scientists realized that the quake might have served to unclamp the locked San Andreas Fault. The media quickly picked up on scientists' concern. U.S. Geological Survey seismologist Allan Lindh told reporters that "the earth has done everything it can to alert the people of southern California that trouble is on the way." Pressed for specifics that the scientific community couldn't provide, Lindh spoke in even stronger terms. "Everyone agrees," he said, "that Los Angeles and San Bernardino should treat this like a final warning. It's like when you clean up camp. It's time to make that last pass through our cities, homes, and lives and act as if the damn thing will happen tomorrow."

Lindh caught some flack in professional circles for what some colleagues perceived as overly alarmist rhetoric. Indeed, over the months and years following Landers the southern San Andreas Fault again remained stubbornly locked. But if, now as then, the best the experts can say is that the Big One could strike tomorrow or thirty years from now, when does one prepare, if not today?

CHAPTER 18

Whither Earthquake Prediction?

> All of us would rather be somewhere else, and this is not
> really the kind of forum which would ideally settle a
> scientific issue. . . . Nevertheless, I think it is clear that
> earthquake prediction is a very special field. We all have
> certain social responsibilities, and I think this is the kind
> of thing we simply must go through.
> —CLARENCE ALLEN, 1981 NEPEC hearing
> on Brady-Spence Prediction

Are earthquakes predictable? The title of this book implies an answer and suggests a paradox. We cannot say it will always be the case, but, given the state of earthquake science at the present time, earthquakes are unpredictable. On a geological time scale they occur like clockwork; on a human time scale they are vexingly, almost determinedly, irregular. The next Big One in California might be next year, or thirty years from now. It might not happen for one hundred years. Or, as seismologists sometimes point out, it could have happened already and the P-wave hasn't gotten here yet. (This would be geek humor.) The odds are that, whenever the San Andreas Fault—or the Hayward, or the New Madrid Seismic Zone, or the North Anatolian Fault—is about to let loose, the only heads up we can hope for is a foreshock or a foreshock sequence. Even then, the odds are that the foreshock sequence won't look any different from the small earthquakes that pop off on these faults on a fairly regular basis.

Aside from foreshocks, earthquake precursors remain elusive at best. At any give time a number of possibilities remain in the running, it seems. At a 2008 meeting of the Seismological Society of America, respected U.S. Geological Survey seismologist Yuehua Zeng presented intriguing results using the Load-Unload Response Ratio (LURR) method. This method, developed in the 1990s by scientists in China, involves the old idea of tidal triggering, but with a twist. According to the theory underlying the method, small tidal stresses don't necessarily trigger big earthquakes, but they do tend to trigger small earthquakes in a region when stress starts to build toward a large quake. That is, as the stress starts to build, the earth's crust is effectively more easily disturbed by small fluctuations in stress—more ticklish, if you will. Thus the correlation between high tidal stresses and small quakes—something we can measure—increases in advance of a big quake. Or so the theory goes. Playing the usual game of looking at data prior to earthquakes that have already happened, a number of researchers, including Zeng and his colleagues, believe they have found intriguing—some dare say promising—results.

Another study, reported in the journal *Nature* in July 2008, used data from sensitive seismometers installed in a deep borehole in Parkfield, California, and found minute changes in wave speeds in the hours before two small earthquakes. The authors concluded that small increases in wave speed resulted from increases in stress in the vicinity of the quakes. The study, which harks back to early Vp/Vs investigations, employed far more sophisticated instrumentation than that available in the 1970s. In the minds of many the limitations of this study also harked back to an earlier era: the limited data, the absence of rigorous statistics. Still, the results remind us that the theory of dilatancy was never dead and buried.

Yet we have been here before, so many times. Looking back at earthquakes that have already happened scientists identify apparently promising precursors. Changes in seismic wave velocities, radon release, accelerating moment release, earthquake chains—the patterns look so convincing, surely they must be real.

Yet for all of the promise, for all of the waves of optimism, no prediction method has panned out when put to the one real test, namely predicting earthquakes that haven't already happened. Small wonder that seismologists now emerge as dyed-in-the-wool skeptics in the face of optimism that has been generated in other fields. We have read the book before. We know how it ends.

Maybe our hard-won skepticism will prove to serve us poorly, blinding us to truly promising results that emerge from other scientific quarters. In his classic book, *The Structure of Scientific Revolutions*, Thomas Kuhn described the barriers to paradigm shifts in science, in particular the tendency of a mainstream community to be tradition bound and hostile to genuinely new ideas. Certainly the charge is made, loudly and often, from other scientific quarters as well as those outside the mainstream community.

And sometimes the charge is made from other quarters. Social scientist Richard Olson portrays the earthquake community's reaction to the Brady-Spence prediction in explicitly Kuhnian terms. In Olson's view the mainstream community ignored Brady far too long, finally going on the attack in something akin to a witch hunt. "Neglect," he writes, "appears to be the first-level response by the scientific community to unorthodoxy. If attention to the novelty is forced, 'passive control' (polite hearing, but no response) is the second step. 'Active control' (threats, sanctions, rewards, attacks) comes later, partly because most incipient challenges fail to survive neglect and passive control." It is a compelling argument, marred only by the fact that sometimes, maybe most of the time, unorthodox science is just bad science.

Whether reliable earthquake prediction will ever be possible, we can't say. But the point bears repeating: for all of the optimism, for all of the past promise, for all of the past hopes generated by apparently promising prediction research, none of it has borne fruit. And as scientists continue to pursue prediction methods, unpredicted quakes continue to strike.

But then there is the paradox: as a physical process, there is a healthy measure of predictability associated with earthquakes. In active regions like California, Alaska, Turkey, and Japan, we know where earthquakes tend to happen, and at what long-term rates. We know that, more generally, a whopping 99 percent of the world's earthquake energy release will happen in the small fraction of the planet that lies along active plate boundaries. Most of this energy will further be released in collision zones, where the seafloor meets and subducts under a continent, or a continent-continent collision such as the ongoing pile-up of India into Eurasia.

We also know that, although we don't fully understand why, earthquakes tend to cluster in time and space. It is rumored to be a Richterism, the observation that "when you get a lot of earthquakes, you get a lot of earthquakes." Recently developed theories of earthquake interactions, focusing on how one earthquake disturbs the crust around it, explain some aspects of clustering. As earlier chapters discussed, it is possible that earthquakes are driven by some external process—anything from tidal stresses to fluid release deep in the crust to magmatic activity—that further contributes to clustering, and to earthquake nucleation. Attempts to develop reliable prediction methods have failed, but that doesn't mean that processes aren't alive and well in the earth.

In any case, whatever the cause, earthquakes do tend to beget other earthquakes. For this reason, it is not a bad bet to predict significant earthquakes after one or more significant earthquakes has already happened. Not such a bad bet, but not terribly satisfying either. But if the more-of-the-same method mostly works, this says that there is a measure of predictability in the system. Some of the methods developed in recent years unconsciously exploit clustering; others, for example parent-daughter methods, take the bull by the horns directly.

This is not earthquake prediction, but it is predictability. And thus we arrive at the heart of the paradox. We can't predict any one earthquake, but, standing back to look at earthquakes as a system, we can say a lot about how the system behaves. And as we continue to learn more about the system, for example to develop better understandings of earthquake-earthquake interactions, presumably we learn more about predictability.

As a young scientist Thomas Jordan worked with Frank Press, and had a front-row seat for the earthquake prediction heyday in the 1970s. He saw the optimism, the promises, and, ultimately, the failures. This experience left him "skeptical of 'silver bullets,'" not to mention wary of bold promises. Working to launch the Center for Earthquake Probability three decades alter, he is gunning neither to develop a reliable

prediction method nor to debunk any proposed method. He is gunning to understand earthquakes, not as earth-shaking events that occur
in isolation, but as the paroxysms of an enormously complicated, fasci-
nating scientific system. "The basic point," he says, "is that the study of
predictability is a main road towards the physical understanding of system behavior."

Every science has its buzzwords. In earthquake science these days
one hears a lot about "systems science." Earthquake scientists have in
recent decades recognized that understanding earthquakes is not a
matter of solving a mathematical equation, or even understanding exactly what happens on a fault when an earthquake occurs. Our job is a
lot harder than that. We have to understand not only faults but also
how they interact with one another. We have to understand not only
the earth's crust but also how the brittle crust interacts with the more
plasticky layers below. We have to understand not only individual
earthquakes, but also how earthquakes interact with other earthquakes.
We have to understand the complex role of fluids in the crust, and
whether migration of fluids can give rise to electromagnetic and other
signals.

Jordan pitches it as a selling point to bright prospective graduate
students, that earthquake science is a field in which the most fundamental problem—reliable earthquake prediction—remains to be
solved. The hitch, which he does not take pains to mention, but which
bright students undoubtedly figure out, is that it might not be solv-
able—at least, not in the sense of developing the kind of reliable, short-
term prediction the public wants. As a sales pitch, this one appeals to a
certain breed of student. The students and researchers who pursue prediction methods are a self-selected group of individuals, individuals
who, as a rule, are not lacking in confidence. Earthquake prediction
research is no place for wimps. If there is any single key to understanding the drama—debates heated to the point of combustion—that
play out over earthquake prediction research within the mainstream
community, this might be it. If there is any single key to understanding how and why prediction research has gone so far awry, this might
be it.

As educated scientists we are quick to condemn the "fools and charlatans"—the armchair seismologists who are convinced they can unlock the most enduring mystery in seismology. What really drives any researcher, trained or otherwise, is a difficult thing to know. The question of motivation aside, whatever drives scientists who are involved with earthquake prediction research, clearly they are not immune to the perils of self-deception to which amateurs are prone. Even well-trained scientists, even brilliant scientists, can fool themselves in their quest to prove something they believe or want to be true. Even well-trained scientists, even brilliant scientists, can fall short in their appreciation of: "one of the most fundamental rules of science—self-criticism."

It is a hard thing for any scientist to do, to admit they have been on a path that isn't going to go anywhere. But to push the edge of the scientific envelope requires the courage of one's convictions. The bigger the challenge to current paradigms, the greater the courage required. If scientists backed away from every research avenue that met with heated resistance from the community, science would never move forward. Thus one can perhaps appreciate that earthquake prediction researchers, the individuals who are a select fraternity in the first place, will not throw in the towel too easily. And given the nature of this fraternity one can perhaps understand if not appreciate that some researchers will throw in the towel far too slowly.

Leading seismologist Hiroo Kanamori is aware that prediction proponents are sometimes not honest actors. At the same time he worries that the seismology community could be throwing the proverbial baby out with the bathwater in its rush to respond to the fervor of faith-based research. In Kanamori's mind there is, or should be, a clear distinction between investigations of the scientific processes that have been linked to earthquake prediction and earthquake prediction research per se. He thinks that the community has a lot to learn about the basic scientific processes that control earthquakes, and that reported observations of anomalous signals, for example groundwater chemistry and electromagnetic signals, may stand to tell us a lot about how earthquakes happen. Like his colleague Tom Jordan, Kanamori's focus is squarely on the science of earthquakes. A better understanding of earthquakes as a physical phenomenon is, in the end, what earthquake scientists are after. And a better understanding might not ever lead to a higher degree of predictability, but it is the one best hope we have.

Is there a path that will lead to reliable, short-term earthquake prediction? Some respected scientists conclude such a path does not exist. Others are convinced that a path exists, and they are on it. The mainstream seismological community remains agnostic at best. Rereading his 1982 article on earthquake prediction a quarter-century after it was published, Clarence Allen observed that "many of the various reasons given for encouragement and discouragement at that time seem just as valid today as twenty-six years ago!" The fact that the field has been running in place all this time leaves Allen feeling less optimistic than he was in 1982; he is convinced that progress will require a strong emphasis on basic research, and that other lines of hazard-reduction efforts should remain higher priorities. Still, the scientist who weighed in with sage wisdom throughout the heyday of the 1970s and the hangover of the 1980s continues to believe that prediction "represents a legitimate research area," and one the mainstream community should not write off as a lost cause.

Indeed, earth scientists, including some seismologists, continue to pursue earthquake prediction research. Like most research the work is mostly carried out behind closed doors; unlike most scientific research the work doesn't always stay behind closed doors. When a biologist develops and tests theories he or she runs experiments in a laboratory. Lab animals might have a direct stake in the investigations as they proceed but humans remain safely removed from the fray until the results are in. Until results are in, the business of science is often far more messy than people tend to imagine; also at times far less subjective. *Science* is the unvarnished pursuit of truth. *Scientists* are human, neither infallible nor entirely immune from the foibles that plague those we rush to decry. But as researchers start to feel their earthquake prediction methods have promise, results and debates will continue to find their way into the scientific literature, and sometimes land with great splash in the public arena.

When the curtain is pulled back, the public gets a glimpse of just how messy science can be. As the NEPEC evaluation of the Brady-Spence prediction came to a close back in 1981, Clarence Allen ended the session with what social scientist and author Richard Olson called a "curious statement." After expressing appreciation for Brady's "stamina and spirit," Allen observed that "all of us would rather be somewhere else, and this is not really the kind of forum which would ideally settle a scientific issue, which we are asked to be critical of our colleagues. Nevertheless I think it is clear that earthquake prediction is a very special field. We all have certain social responsibilities, and I think this is the kind of thing we simply must go through."

The seismologist is left to wonder what could be considered "curious" about the statement. It speaks directly to the challenges we face as scientists when active research generates keen interest outside the halls of science. It speaks directly to the frustrations but also the sense of responsibility we feel, working to understand the science of earthquakes and sometimes to weigh in on matters of great societal concern, in the midst of enormous uncertainties and highly charged debates.

Scientists understand how messy science can be. Scientists know that science is sometimes a contact sport. Scientists understand that as any science strives to push the frontiers of knowledge, not every path will take us in the right direction. This is the nature of research. For better and sometimes for worse, where earthquake prediction is concerned, we—everyone who worries about the sometimes violent paroxysms of our amazing dynamic planet—are all in the middle of it.

Acknowledgments

One of the best things about writing science books and articles for a nonspecialist audience is the opportunity it gives me to talk to my colleagues about matters we would otherwise never discuss. I am indebted, as always, to a long list of colleagues—many of whom I am happy to count as friends: Bob Geller, Chris Scholz, Friedemann Freund, Bob Castle, Nancy King, Jim Savage, Bob Dollar, Peter Molnar, Tom Jordan, Jeremy Zechar, Ruth Harris, Jim Lienkaemper, Robin Adams, Kelin Wang, David Hill, Dave Jackson, Ross Stein, Rob Wesson, Jim Rice, Mike Blanpied, Vladimir Keilis-Borok, Andy Michael, Jeanne Hardebeck, Arch Johnston, Ken Hudnut, Karen Felzer, Tom Heaton, Mark Zoback, Seth Stein, Lloyd Cluff, Jeremy Thomas, Malcolm Johnston, John Filson, and Tom Hanks. I am further indebted to Don Anderson for passing along newspaper clippings, and to Clay Hamilton, Kelin Wang, and Roger Bilham for making figures available.

I owe a special debt of gratitude to colleagues who read early (read: rough) drafts of the manuscript, and provided helpful feedback: Max Wyss, John Vidale, Hiroo Kanamori, Roger Bilham, and Clarence Allen. In the course of my research for the book I realized that Clarence Allen is kind of the Forrest Gump of the earthquake prediction scene, except that it was not by accident that he was the man on the spot at so many pivotal moments. Beneath Clarence's modest and unassuming manner lurks a scientific acumen and intellect that has been valued and tapped by his colleagues and the community throughout his long career.

To my mind, at least, this is a story that has no villains, but three heroes emerge. Clarence Allen is one; the second is David Bowman, whose balance of determination and intellectual integrity represents the best face of science. The third is John Filson, who took the helm of the USGS Office of Earthquake Studies and immediately found himself at ground zero of a maelstrom, one he navigated with consummate insight and grace.

I am further indebted to a number of individuals outside the mainstream scientific community who have been willing to engage in a dialog: in particular Petra Challus, Don Eck, Zhonghou Shou, and Brian Vanderkolk (known in earthquake prediction circles as Skywise). I am not sure they will be entirely happy with what I have to say. I hope they, and others, will see the book as critical but fair; also balanced, as evidenced by the strong likelihood that some of my colleagues will not be happy with what I have to say, either. Science is a contact sport. Anyone who believes that scientists treat one another with kid gloves and outsiders with brickbats has never had a front-row seat for the peer review process.

Certainly Brian Brady can attest to this. His exploration of unconventional theories, and his quest to find insights from apparently disparate branches of science, leads to interesting ideas. In the end I am forced to agree with the members of the National Earthquake Prediction Evaluation Council (NEPEC) who concluded that Brady's arguments were problematic in a number of respects and his theories not formulated with sufficient rigor to justify a prediction. Whether there are, as he suggests, insights to earthquake processes that can be drawn from critical-point theory, the jury is still out.

As always I am thankful to the many talented individuals who helped turn a manuscript with rough edges into such a lovely book: editor Ingrid Gnerlich, copyeditor Dawn Hall, production editor Sara Lerner, assistant editor Adithi Kasturirangan, and managing editor Elizabeth Byrd. I am further grateful to the Princeton design team for artistic wizardry with the cover design, and to the marketing team for helping this book find its way in the world.

Last but not least, a sizable measure of gratitude is reserved for my colleague and friend, Greg Beroza. It was about five years ago that I stood in his office and he remarked, "What we need is a good general book on earthquake prediction. Why don't you write one?" So, really, this whole silly thing is his fault. Thank you, Greg.

Notes

Chapter 1. Ready to Rumble

Page

1–2. Robert Dollar, personal communication, 2008.

4. Nancy King, personal communication, 2009.

4. Robert Dollar, personal communication, 2008.

6. Drudge Report, http://www.drudgereport.com, June 22, 2006.

6. Y. Fialko, "Interseismic Strain Accumulation and the Earthquake Potential of the Southern San Andreas Fault," *Nature* 441 (2006): 968–71.

7. Joel Achenbach, *Washington Post*, January 30, 2005.

10. N. E. King et al., "Space Geodetic Observation of Expansion of the San Gabriel Valley, California, Aquifer System, during Heavy Rainfall in Winter, 2004–2005," *Journal of Geophysical Research* 112 (March 2007), doi:10.1029/2006JB004448.

10. "Valley a Victim of Battle of the Bulge," *Pasadena Star-News*, April 10, 2007.

10. "Scientists Intrigued by Quake Forecasts," *Los Angeles Times*, April 18, 2004, B1.

Chapter 2. Ready to Explode

Page

12. Judith R. Goodstein, *Millikan's School* (New York: W. W. Norton, 1991); Susan E. Hough, *Richter's Scale: Measure of an Earthquake, Measure of a Man* (Princeton, NJ: Princeton University Press, 2006).

12–13. Ben M. Page, Siemon W. Muller, and Lydik S. Jacobsen, "Memorial Resolution, Bailey Willis (d. 1949)," http://histsoc.stanford.edu/pdfmem/WillisB.pdf.

13–14. Carl-Henry Geschwind, *California Earthquakes: Science, Risk, and Politics* (Baltimore: Johns Hopkins University Press, 2001).

15. Harold W. Fairbanks, "Pajaro River to the North End of the Colorado Desert," in *The California Earthquake of April 18, 1906: Report of the State Earthquake Investigation Commission*, 2 vols. and atlas (Washington, DC: Carnegie Institution of Washington, 1908–10).

15, 17. Judith Goodstein, Millikan's School.

17. Bailey Willis, "A Fault Map of California," *Bulletin of the Seismological Society of America* 13 (1923): 1–12.

18. Tachu Naito, "Earthquake-Proof Construction," *Bulletin of the Seismological Society of America* 17 (1927): 57–94.

18–19. Page, Muller, and Jacobsen, "Memorial Resolution, Bailey Willis."

20. Bailey Willis, *A Yanqui in Patagonia* (Stanford, CA: Stanford University Press, 1947).

20. Geschwind, *California Earthquakes: Science, Risk, and Politics*.

22. "Prof. Willis Predicts Los Angeles Tremors," *New York Times*, November 4, 1925.

22. "General Earthquake or Series Expected," *Sheboygan Press*, November 16, 1925, 5.

22. "Faux Pas," *Time* magazine, November 16, 1925.

24. Robert T. Hill, *Southern California Geology and Los Angeles Earthquakes* (Los Angeles: Southern California Academy of Sciences, 1928).

24. Clarence Allen, personal communication, 2008.

25. "It is generally believed that Dr. Willis' service": "Science's Business," *Time* magazine, February 27, 1928.

25–26. "Eyewitness Accounts of Cataclysm," *Oakland Tribune*, March 11, 1933.

26. Charles F. Richter, *Elementary Seismology* (San Francisco: W. H. Freeman, 1958).

28. International News Service Press Release, Stanford University, March 1933.

28. "The Wasatch Fault . . .": "Salt Lake City in Earthquake Zone," *Woodland Daily Democrat*, Thursday, July 20, 1933.

Chapter 3. Irregular Clocks

Page

29. Pecora quote: "Californians Getting Jumping Nerves—Earthquakes," *Las Vegas Daily Optic*, March 31, 1969.

29. Grove Karl Gilbert, "Earthquake Forecasts," *Science* 29 (1909): 121–28.

35. Working Group on California Earthquake Probabilities, Probabilities of Large Earthquakes Occurring in California on the San Andreas Fault, U.S. Geological Survey Open-File Report 88-398, 1988

36–37. Http://www.forttejon.org/historycw.html (last accessed 2/8/2009).

38. "Most of us have an awful feeling":"The BIG ONES," *Stars and Stripes*, August 1, 1992.

38. "I think we're closer than 30 years": ibid.

Chapter 4. The Hayward Fault

Page

39. "Loss of Life—Panic of the People—Full Particulars of Its Effects on the City—Its Effects in Oakland, San Leandro, and Other Places across the Bay—etc., etc.," *San Francisco Morning Call*, October 22, 1868.

42. Ellen Yu and Paul Segall, "Slice in the 1868 Hayward Earthquake from the Analysis of Historical Triangulation Data," *Journal of Geophysical Research* 101 (1996): 16101–118.

43. James J. Lienkaemper and Patrick L. Williams, "A Record of Large Earthquakes on the Southern Hayward Fault for the Past 1800 Years," *Bulletin of the Seismological Society of America* 97 (2007): 1803–19.

46. "Do the right thing": "Seismologists Warn of Looming Quake on Hayward Fault," *Berkeley Daily Planet*, December 18, 2007.

46. James Lienkaemper, personal communication, 2008.

Chapter 5. Predicting the Unpredictable

Page

47. *Eeyore, Be Happy!* A Little Golden Book (Racine, WI: Golden Books Publishing, 1991).

48. Robert M. Nadeau and Thomas V. McEvilly, "Fault Slip Rates at Depth from Recurrence Intervals of Repeating Microearthquakes," *Science* 285 (1999): 718–21.

48. Naoki Uchida et al., "Source Parameters of a M4.8 and Its Accompanying Repeating Earthquakes off Kamaishi, NE Japan: Implications for the

Hierarchical Structure of Asperities and Earthquake Cycle," *Geophysical Research Letters* 34 (1967), doi:10.1029/GL031263.

48. Jeffrey J. McGuire, Margaret S. Boettcher, and Thomas H. Jordan, "Foreshock Sequences and Short-Term Earthquake Predictability on East Pacific Rise Transform Faults," *Nature* 434 (2005): 457–61.

48. R. P. Denlinger, and Charles G. Bufe, "Reservoir Conditions Related to Induced Seismicity at the Geysers Steam Reservoir, Northern California," *Bulletin of the Seismological Society of America* 72 (1982): 1317–27.

49. J. H. Healy, W. W. Rubey, D. T. Griggs, and C. B. Raleigh, "The Denver Earthquakes," *Science* 27 (1968): 1301–10.

49. Harsh K. Gupta et al., "A Study of the Koyna Earthquake of December 10, 1967," *Bulletin of the Seismological Society of America* 59 (1969): 1149–62.

50. W. H. Bakun and T. V. McEvilly, "Recurrence Models and Parkfield, California, Earthquakes," *Journal Geophysical Research* 89 (1984): 3051–58.

52. Roger Bilham, personal communication, 2008.

52. Allan Lindh, "Success and Failure at Parkfield," *Seismological Research Letters* 76 (2005): 3–6.

54. Special issue on the Great Sumatra–Andaman earthquake, *Science* 308 (2005): 1073–208.

55. H. Ishii and T. Kato, "Detectabilities of Earthquake Precursors Using GPS, EDM, and Strain Meters, with Special Reference to the 1923 Kanto Earthquake," *Journal of the Geodesy Society of Japan* 35 (1989): 75–83.

55. M. Wyss and D. C. Booth, "The IASPEI Procedure for the Evaluation of Earthquake Precursors," *Geophysical Journal International* 131 (1997): 423–24.

55. Hiroo Kanamori, "Earthquake Prediction: An Overview," *IASPEI* 81B, (2003): 1205–16.

55. M. Wyss, "Nomination of Precursory Seismic Quiescence as a Significant Precursor," *Pure and Applied Geophysics* 149 (1997): 79–114.

55–56. "Too bad, but they don't. Roughly, little shocks on little faults, all over the map, any time; big shocks on big faults": Richter, memo to Bob Sharp, 1952, Box 23.6. Papers of Charles F. Richter, 1839–1984, California Institute of Technology Archives.

56. W. Inouye, "On the Seismicity in the Epicentral Region and Its Neighbourhood before the Niigata Earthquake," *Quarterly Journal of Seismology* 29 (1965): 139–44.

56. K. Mogi, "Some Features of Recent Seismic Activity in and near Japan

(2). Activity before and after Great Earthquakes," *Bulletin of the Earthquake Research Institute of Japan* (1969).

56. *Eeyore, Be Happy!*

57. Ruth Simon, abstract, *Eos, Transactions, American Geophysical Union* 759 (1976): 59.

57. Earthquake Information Bulletin 10, March–April, 1978.

57. "Swedish Cows Make Lousy Earthquake Detectors: Study," *The Local*, January 13, 2009 (http://www.thelocal.se/16876/20090113/, last accessed 2/9/2009).

57. "The Predicted Chinese Quakes," *San Francisco Chronicle*, July 30, 1976, 34.

Chapter 6. The Road to Haicheng

Page

58. Kelin Wang et al., "Predicting the 1975 Haicheng Earthquake," *Bulletin of the Seismological Society of America* 96 (2006): 757–95.

58–62. Susan E. Hough, "Seismology and the International Geophysical Year," *Seismological Research Letters* 79 (2008): 226–33.

62. Ad hoc panel on earthquake prediction, "Earthquake Prediction: A Proposal for a Ten-Year Program of Research," Office of Science and Technology, Washington DC, September 1965.

63. "Big Quake Center Will Be at Menlo Park," *San Francisco Chronicle*, October 8, 1965.

63. "Interagency squabbling . . .": Geschwind, *California Earthquakes: Science, Risk, and Politics*, 200.

63. "Having acquired a reputation . . .": ibid., 143.

64. Reuben Greenspan, "Earthshaking Prediction," *Pasadena Star News*, December 21, 1972.

64. "Charlatans, fakes, or liars": "Experts Don't Agree on Quake Prediction," *Long Beach Press-Telegram*, February 15, 1971, A3.

65. D. Anderson quote in ibid.

65. "If you had asked me . . .": "Daily Tremor Forecasts May Be Available Soon," *Oakland Tribune*, June 9, 1971, E9.

66. George Alexander, "New Technology May Bring Quake Forecasts," *Los Angeles Times*, February 22, 1973, part 2, page 1.

66–68. Peter Molnar, personal communication, 2008.

68. Rob Wesson, personal communication, 2008.

68–69. Christopher H. Scholz, Lynn R. Sykes, and Yash P. Aggarwal, "Earthquake Prediction: A Physical Basis," *Science* 31 (1973): 181, 803–10.

70–71. Robin Adams, personal communication, 2007.

71–84. Wang et al., "Predicting the 1975 Haicheng Earthquake," 757–95.

72. Ross S. Stein, "The Role of Stress Transfer in Earthquake Occurrence," *Nature* 402 (1999): 605–9.

78. "China Quake Severe," *Wisconsin State Journal*, March 14, 1975, section 4, page 8.

83. Helmut Tributsch, *When the Snakes Awake* (Cambridge, MA: MIT Press, 1982).

83–84. Kelin Wang, personal communication, 2008.

84. Kelin Wang, personal communication, 2008.

84. "Quake Split Hotel in 2, Visiting Australian Says," *Los Angles Times*, July 28, 1976, 1.

Chapter 7. Percolation

Page

86. M. Manga and C-Y. Wang, "Earthquake Hydrology," IASPEI volume, 1991.

86–87. N. Yoshida, T. Okusawa, and H. Tsukahara, "Origin of Deep Matsushiro Earthquake Swarm Fluid Inferred from Isotope Ratios," *Zisin* 55 (2002): 207–16 (in Japanese with English abstract).

88. Robert M. Nadeau and David Dolenc, "Nonvolcanic Tremors beneath the San Andreas Fault," *Science* 307 (2005): 389; K. Obara, "Nonvolcanic Deep Tremor Associated with Subduction in Southwest Japan," *Science* 296 (2002): 1679–81.

88. David R. Shelly, Gregory C. Beroza, and S. Ide, "Non-Volcanic Tremor and Low-Frequency Earthquake Swarms," *Nature* 446 (2007): 305–7.

88–89. David P. Hill, John O. Langbein, and Stephanie Prejean, "Relations between Seismicity and Deformation during Unrest in Long Valley Caldera, California, from 1995 to 1999," *Journal of Volcanology and Geothermal Research* 127 (2003): 175–93.

90. David Hill, personal communication, 1998.

91. N. I. Pavlenkova, "The Kola Superdeep Drillhole and the Nature of Seismic Boundaries," *Terra Nova* 4 (2007): 117–23.

92. "Scenarios of this kind . . .": Manga and Wang, "Earthquake Hydrology."

92–93. Dapeng Zhao et al., "Tomography of the Source Are of the 1995 Kobe Earthquake: Evidence for Fluids at the Hypocenter?" *Science* 274 (1996): 1891–94.

93–94. Egill Hauksson and J. G. Goddard, "Radon Earthquake Precursor Studies in Iceland," *Journal of Geophysical Research* 86 (1981): 7037–54.

94. George Alexander, "Possible New Quake Signals Studied," *Los Angeles Times*, October 27, 1979, 1, 28.

94. George Alexander, "Quake That Hit Without Warning Puzzles Scientists," *Los Angeles Times*, October 27, 1979, 3.

94. Sergey Alexander Pulinets, "Natural Radioactivity, Earthquakes, and the Ionosphere," *Eos, Transactions, American Geophysical Union* (2007), doi:10 .1029/2007EO200001.

95. Hiroo Kanamori, personal communication, 2008.

Chapter 8. The Heyday

Page

96. George Alexander, "Can We Predict the Coming California Quake?" *Popular Science*, November, 1976, 79–82.

96. "Recent results . . .": Frank Press, "U.S. Lags in Quake Prediction," *Xenia Daily Gazette*, February 27, 1975.

96. "Seismologists are getting close . . .": *Los Angeles Times*, April 27, 1975.

97. R. Hamilton, "Chinese Credited with Prediction of Earthquake," *Bridgeport Sunday Post*, November 9, 1975.

97. "As early as . . .": "Geologists Zero in on Accurate Earthquake Predictions," *Fresno Bee*, November 9, 1975.

98–99. Robert O. Castle, John N. Alt, James C. Savage, and Emery I. Balazs, "Elevation Changes Preceding the San Fernando Earthquake of February 9, 1971," *Geology* 5 (1974): 61–66.

98. "As conservative as they come . . .": Robert Castle, personal communication, 2008.

99. Max Wyss, "Interpretation of the Southern California Uplift in Terms of the Dilatancy Hypothesis," *Nature* 266 (1977): 805–8.

99. "You bet he was.": Robert Castle, personal communication, 2008.

100. Geschwind, *California Earthquakes: Science, Risk, and Politics*, 204–5.

100. Robert O. Castle, Jack P. Church, and Michael R. Elliott, "Aseismic Uplift in Southern California, *Science* 192 (1976): 251–53.

100. "I think there's reason to be concerned . . .": "'Bulge' on Quake Fault 'May Be Message,'" *Long Beach Independent Press-Telegram*, March 18, 1976, A3.

100. "Predictable headlines followed": ibid.

101. Frank Press, "A Tale of Two Cities," annual meeting, American Geophysical Union, 1976.

101. "Quake Prediction Saved Chinese," *Capital Times*, April 15, 1976.

101. "Earthquake prediction, long treated as the seismological family's weird uncle, has in the last few years become everyone's favorite nephew": George Alexander, *Popular Science*, November, 1976, 79–82.

101–102. George Alexander, "Scientist's Prediction of Quake Comes True," *Los Angeles Times*, April 11, 1974, 1.

102. George Alexander, "In His Own Words: Just like the Movie, a Young Scientist Predicts an L.A. Earthquake," *People* magazine, May 17, 1976, 49–55.

102–104. Geschwind, *California Earthquakes: Science, Risk, and Politics*.

102. Federal Council for Science and Technology, Ad Hoc Interagency Working Group for Earthquake Research, William T. Pecora, Proposal for a Ten-Year National Earthquake Hazards Program; A Partnership of Science and the Community. Prepared for the Office of Science and Technology and the Federal Council for Science and Technology, 1968–69. Earthquake Prediction and Hazard Mitigation Options for USGS and NSF Programs, National Science Foundation, Research Applications Directorate (RANN) and USGS, 1976.

102–103. Task Force on Earthquake Hazard Reduction, Karl V. Steinbrugge, Chair, Earthquake Hazard Reduction, Office of Science and Technology, 1970.

104. Earthquake Prediction and Hazard Mitigation Options for USGS and NSF Programs, National Science Foundation, Research Applications Directorate (RANN), and USGS, 1976.

104. "Time to get busy . . .": Wallace, Earthquakes, minerals, and me.

104. "Earthquake in Guatemala," *National Geographic*, June 1976, 810–29; "Can We Predict Earthquakes," *National Geographic*, June 1976, 830–35.

104. "There is much to learn . . .": The Guatemalan Earthquake of

2/4/1976, a Preliminary Report, U.S. Geological Survey Professional Paper 1002, Washington DC: U.S. Government Printing Office.

105. Lloyd Cluff, personal communication, 2008

105. "Earthquake prediction failed . . .": Earthquake Hazards Act Fails in House," *Northwest Arkansas Times*, September 29, 1976, 27.

105–106. "Scientist Urges World Quake Study," *Pasadena Star-News*, February 24, 1977, 1.

106. "Earthquake hazard mitigation become entrenched . . .": Geschwind, *California Earthquakes: Science, Risk, and Politics*, 212.

106. Christopher H. Scholz, "Whatever Happened to Earthquake Prediction?" *Geotimes* 17 (March 1997).

106. Conrad, editorial cartoon, *Los Angeles Times*, April 23, 1976.

107. Peter Molnar, personal communication, 2003.

107. Lloyd Cluff, personal communication, 2008.

107. Paul Houston, "Funds for Quake Research Killed," *Los Angeles Times*, March 21, 1976, part 2, page 7.

Chapter 9. The Hangover

Page

108. C. R. Allen, "Earthquake Prediction—1982 Overview," *Bulletin of the Seismological Society of America* 72 (1982): S331–S335.

108. At http://www.nehrp.gov/about/.

109. George Alexander, "Rock-Layer Changes May Bring on Quake Advisories," *Los Angeles Times*, January 27, 1980, Metro, 1.

109. "California Tenses over Fault Indicators," *Pacific Stars and Stripes*, August 29, 1980, 8.

109. George Alexander, "Bubbling Gases concern Earthquake Researchers," *Los Angeles Times*, October 8, 1981, 1, 17; George Alexander, "Possible Quake Precursor Observed over Wide Area," *Los Angeles Times*, October 17, 1981, 1, 22.

110. Robert Reilinger and Larry Brown, "Neotectonic Deformation, Near-Surface Movements and Systematic Errors in U.S. Releveling Measurements: Implications for Earthquake Prediction," *Earthquake Prediction, An International Review*, American Geophysical Union Maurice Ewing Monograph Series 4 (1981): 422–40; David D. Jackson, Wook B. Lee, and Chi-Ching Liu,

"Height Dependent Errors in Southern California Leveling," *Earthquake Prediction, An International Review*, American Geophysical Union Maurice Ewing Monograph Series 4 (1981): 457–72.

110. Ross Stein, personal communication, 2005

110. David Jackson, personal communication, 2005.

110. R. Kerr, "Palmdale Bulge Doubts Now Taken Seriously," *Science* 214 (1981): 1331–33.

110. "Clue to Quake Forecasting Sought in Tiny Temblors," *New York Times*, August 11, 1973, 21, 38.

110. "Predicting the Quake," *Time* magazine, August 27, 1973.

111. Chris Scholz, personal communication, 2008

111-112. Richard Stuart Olson, Bruno Podesta, and Joanne M. Nigg, *The Politics of Earthquake Prediction* (Princeton, NJ: Princeton University Press, 1989).

112. Brian T. Brady, "Seismic Precursors before Rock Failures in Mines," *Nature* 252 (1974): 549–52; Brady, "Theory of Earthquakes," *Pure Applied Geophysics* 112 (1974): 701–25; Brady, "Dynamics of Fault Growth: A Physical Basis for Aftershock Sequences," *Pure Applied Geophysics* 114 (1976): 727–39; Brady, "Theory of Earthquakes—IV. General Implications for Earthquake Prediction," *Pure Applied Geophysics* 114 (1976): 1420–36; Brady, "Anomalous Seismicity Prior to Rock Bursts: Implications for Earthquake Prediction," *Pure Applied Geophysics* 115 (1977): 357–74.

112. "How a fault gets there . . .": Brady quote in Olson, Podesta, and Nigg, *Politics of Earthquake Prediction*, 137.

114. "Didn't add to the prediction's credibility": Clarence Allen, personal communication, 2008.

114. Roger Hanson, internal memo to John Filson, July 16, 1980.

114. William Spence and Lou C. Pakiser, Conference Report: Toward Earthquake Prediction on the Global Scale, *Eos, Transactions, American Geophysical Union* 59 (1978): 36–42.

115. Olsen et al., *Politics of Earthquake Prediction*.

115. Clarence Allen, Transcript of Proceedings, *In the Matter Of: The National Earthquake Prediction Evaluation Council Meeting to Receive Evidence and to Assess the Validity of a Prediction Calling for a Great Earthquake off the Coat [sic] of Peru in August 1981* (Reston, VA: USGS, 1981).

115–116. "Quake Prediction Causes Own Shakes," *Science News* 5 (July 4, 1981).

116. "There has been a request . . ." NEPEC transcript, 71.

116. "This isn't a criticism . . ." NEPEC transcript, 74.

116. David Hill, personal communication, 2008.

116. Keiiti Aki, "A Probabilistic Synthesis of Precursory Phenomena," in *Earthquake Prediction: An International Review*. Maurice Ewing Series No. 4., ed. D. W. Simpson and P. G. Richards (Washington DC: American Geophysical Union, 1981): 566–74.

117. Rob Wesson, personal communication, 2008.

117. Brian T. Brady, "A Thermodynamic Basis for Static and Dynamic Scaling Laws in the Design of Structures in Rock," Proceedings of the 1st North American Rock Mechanics Symposium, University of Texas at Austin, June 1–3, 1994, 481–85.

117. "Scientist Stays with Prophecy of Peru Quake," *Rocky Mountain News*, January 27, 1981, 3.

118. "Ace of H . . .": NEPEC transcript, 1981, 245–46.

118. "Fracture has been studied . . .": ibid., 250.

119. "It has a lot of things . . .": ibid., 286.

119. "I guess that is why . . .": ibid.

119. "Previous public work . . .": ibid., 338.

119. "The seismicity patterns . . .": ibid., 339.

119. Conclusion, ibid.

119. Robert Wesson, letter to Mr. Fournier d'Albe, April 17, 1980.

120. Krumpe memo: see Olson, Podesta, and Nigg, *The Politics of Earthquake Prediction*, 125.

120. Clarence Allen, letter to Peter McPherson, July 10, 1981.

120. Clarence Allen, personal communication, 2008.

120. Brian Brady, personal communication, 2008

120–121. See discussion in Olson, Podesta, and Nigg, *The Politics of Earthquake Prediction*, 187.

121. John Filson, report on visit to Lima, 1981.

121. John Filson, personal communication, 2008.

121. "NO PASO NADA," *Expreso*, June 28, 1991, 1.

121. "If he is allowed . . .": John Filson, Peru trip report.

122. John Filson, personal communication, 2008.

122. Mark Zoback, personal communication, 2008.

122–123. Allen, "Earthquake Prediction—1982 Overview," S331–S335.

123. "Quake Readiness: New Solution Cuts across the Fault Lines," *San Francisco Chronicle-Telegram*, July 11, 1982, Sunday Scene.

124. Michael Blanpied, personal communication, 2008.

Chapter 10. Highly Charged Debates

Page

125. Friedemann Freund, "Rocks That Crackle and Sparkle and Glow: Strange Pre-Earthquake Phenomena," *Journal of Scientific Exploration* 17 (2003): 37–71.

125. K. Varotsos, K. Alexopoulos, K. Nomicos, and M. Lazaridou, "Earthquake Prediction and Electric Signals," *Nature* 322 (1986): 120.

127. Francesco Mulargia and Paolo Gasperini, "VAN: Candidacy and Validation with the Latest Laws of the Game," *Geophysical Research Letters* 23 (1996): 1327–30.

127. Max Wyss, personal communication, 2008.

127. Hiroo Kanamori, personal communication, 2008.

127–129. Robert Geller, personal communication, 2007.

129. "At this moment . . .": "Japan Holds Firm to Shaky Science," *Science* 264 (1994): 1656–58.

130–131. Mohsen Ghafory-Ashtiany, personal communication, 2008.

132. Richard A. Kerr, "Loma Prieta Quake Unsettles Geophysicists," *Science* 259 (1989): 1657.

132. Anthony Fraser-Smith, "Low-Frequency Magnetic Field Measurements Near the Epicenter of the Ms7.1 Loma Prieta Earthquake," *Science* 17 (1989): 12.

132. Robert F. Service, "Hopes Fade for Earthquake Prediction," *Science* 264 (1994): 1657.

132-133. Jeremy N. Thomas et al., abstract 4036, General Assembly of the International Union of Geodesy and Geophysics, Perugia, 2007.

133. Malcolm Johnston et al., General Assembly of the International Union of Geodesy and Geophysics, Perugia, 2007.

133-135. Freund, "Rocks That Crackle and Sparkle and Glow," 37–71.

134–135. Helmut Tributsch, *When the Snakes Awake*.

137. Johnston et al., General Assembly of the International Union of Geodesy and Geophysics, Perugia, 2007.

137. Richard Dixon Oldham, "Report on the Great Earthquake of 12 June 1897." *Memoirs of the Geological Society of India* 29 (1899); Roger Bilham, "Tom La Touche and the Great Assam Earthquake of 12 June 1897: Letters from the Epicenter," *Seismological Research Letters* 79 (2008): 426–37.

140. Friedemann Freund, "Cracking the Code of Pre-Earthquake Signals," at http://www.seti.org/news/features/, September 20, 2005.

140. F. Nemec, O. Santolik, M. Parrot, and J. J. Berthelier, "Spacecraft Observations of Electromagnetic Perturbations Connected with Seismic Activity," *Geophysical Research Letters*, 25, doi:10.1029/2007GL, 032517.

Chapter 11. Reading the Tea Leaves

Page

141. Lewis Carroll, *Alice's Adventures in Wonderland* (London: Macmillan, 1865).

142. Mogi, "Some Features of Recent Seismic Activity in and near Japan (2). Activity before and after Great Earthquakes."

142. V. I. Keilis-Borok and V. G. Kossobokov, "Periods of High Probability of Occurrence of the World's Strongest Earthquakes," translated by Allerton Press, *Computational Seismology* 19 (1987): 45–53.

143. R. A. Harris, "Forecasts of the 1989 Loma Prieta, California, Earthquake," *Bulletin of the Seismological Society of America* 88 (1998): 898–916.

144. Chris Scholz, personal communication, 2008.

144–145. V. I. Keilis-Borok et al., "Reverse Tracing of Short-Term Earthquake Precursors," *Physics Earth and Planetary Interiors* 145 (2004): 75–85.

146. V. I. Keilis-Borok, P. Shebalin, K. Aki, A. Jin, A. Gabrielov, D. Turcotte, Z. Liu, and I. Zaliapin, "Documented Prediction of the San Simeon Earthquake 6 Months in Advance: Premonitory Change of Seismicity, Tectonic Setting, Physical Mechanism," abstract, Annual Meeting, Seismological Society of America, Palm Springs, California, 2004.

147. "Science Is Left a Bit Rattled by the Quake That Didn't Come," *Los Angeles Times*, September 8, 2004, A1.

147–148. Jeremy D. Zechar, PhD thesis, "Methods for Evaluating Earthquake Predictions," University of Southern California, 134 pp., 2008.

148. V. P. Shebalin, personal communication, 2008.

Chapter 12. Accelerating Moment Release

Page

150. John Maynard Keynes, "Keynes Took Alf Garnett View on Race," *The Independent* (London), January 31, 1997.

150. Keiiti Aki, "Generation and Propagation of G Waves from the Niigata Earthquake of June 16, 1964: Part 2. Estimation of Earthquake Moment, Re-

leased Energy, and Stress Drop from the G Wave Spectra," *Bulletin of the Earthquake Research Institute of Tokyo* 44 (1965): 237–39.

150. Charles F. Richter, "An Instrumental Earthquake Magnitude Scale," *Bulletin of the Seismological Society of America* 25 (1935): 1–32.

151. David J. Varnes, Proceedings of the 8th Southeast Asian Geotechnical Conference, 2, Hong Kong, 1982, pp. 107–30; Lynn R. Sykes and Steven C. Jaume, "Seismic Activity on Neighboring Faults as a Long-Term Precursor to Large Earthquakes in the San Francisco Bay Area," *Nature* 348 (1990): 595–99; Charles G. Bufe and David J. Varnes, "Predictive Modeling of the Seismic Cycle of the Greater San Francisco Bay Region," *Journal of Geophysical Research* 98 (1993): 9871–83; David D. Bowman and Geoffrey C. P. King, "Accelerating Seismicity and Stress Accumulation before Large Earthquakes," *Geophysical Research Letters* 28 (2001): 4039–42.

151. Shamita Das and Christopher Scholz, "Off-Fault Aftershock Clusters Caused by Shear Stress Increase?" *Bulletin of the Seismological Society of America* 71 (1981): 1669–75.

151. Geoffrey King, Ross S. Stein, and J. Lin, "Static Stress Changes and the Triggering of Earthquakes," *Bulletin of the Seismological Society of America* 84 (1994): 935–53.

153. David D. Bowman and Geoffrey C. P. King, "Stress Transfer and Seismicity Changes before Large Earthquakes," *Comptes Rendus de l'Academie des Sciences, Sciences de la terre et des planets* 333 (2001): 591–99.

154. Andrew J. Michael, "The Evaluation of VLF Guided Waves as Possible Earthquake Precursors," U.S. Geological Survey Open-File Report 96-97.

154–155. Andrew Michael, personal communication, 2008.

156. Andrew J. Michael, Jeanne L. Hardebeck, and Karen R. Felzer, "Precursory Acceleration Moment Release: An Artifact of Data-Selection?" *Journal of Geophysical Research* (2008).

157. David Bowman, personal communication, 2008.

Chapter 13. On the Fringe

Page

158. "What ails them . . .": Charles F. Richter, memo, 1976, Box 26.17, Papers of Charles F. Richter, 1839–1984, California Institute of Technology Archives.

158–159. Tecumseh's Prophecy; preparing for the next New Madrid earthquake; a plan for an intensified study of the New Madrid seismic zone, U.S. Geological Survey Circular 1066, Robert M. Hamilton, ed., 1990.

159. Papers of Charles F. Richter, 1839–1984, California Institute of Technology Archives.

160. John R. Gribbin and Stephan H. Plagemann, *The Jupiter Effect: The Planets as Triggers of Devastating Earthquakes* (New York: Vintage Books, 1976). George Alexander, "Big LA Quake in '82? Experts Not Shaken by Theory," *Los Angeles Times*, September 13, 1974, part 2, page 1.

160. "There is a distinct tendency for the number of small earthquakes in southern California to increase slightly at the end of the summer . . .": Richter, letter to Harry Plant, March 20, 1958, Box 30.3, Papers of Charles F. Richter, 1839–1984, California Institute of Technology Archives.

160. John R. Gribbin and Stephen H. Plagemann, *The Jupiter Effect: The Planets as Triggers of Devastating Earthquakes* (New York: Vintage Books, 1976).

160. George Alexander, "Big LA Quake in '82? Experts Not Shaken by Theory," *Los Angeles Times*, September 13, 1974, part 2, page 1.

161. Jeffrey Goodman, *We Are the Earthquake Generation* (New York: Berkley Publishing, 1978).

161–162. Earthquake Information Bulletin 10, March–April 1978.

162–163. Gordon-Michael Scallion, "Earth Changes Report," 1992.

163–164. Zhonghao Shou, "Earthquake Clouds: A Reliable Precursor" (in Turkish), *Science and Utopia* 64 (1999): 53–57.

164. Petra Challus and Don Eck at http://www.quakecentralforecasting .com/.

164. Brian Vanderkolk (aka Skywise), at http://www.skywise711.com/ quakes/EQDB/ (last accessed 2/9/2009).

165. Brian Vanderkolk, personal communication, 2009.

166. "A few such persons . . .": Charles F. Richter, , memo, 1976, Box 26.17, Papers of Charles F. Richter, 1839–1984, California Institute of Technology Archives.

168. Robert Geller, personal communication.

168. Roger Bilham et al., "Seismic Hazard in Karachi, Pakistan: Uncertain Past, Uncertain Future," *Seismological Research Letters* 78 (2007): 601–13.

169. R. Mallet, *Great Neapolitan Earthquake of 1857: The First Principles of Observational Seismology*, vol. 1. (London: Chapman and Hall, 1862), 172.

Chapter 14. Complicity

Page

171. "An incensed Charles Richter . . .": Geschwind, *California Earthquakes: Science, Risk, and Politics*, 146.

171. "Californians Getting Jumping Nerves—Earthquakes," *Las Vegas Daily Optic*, March 31, 1969, 2.

172. "Slid? Slud? Slood?—Old Winery Straddles Quake Line," *Fresno Bee*, April 27, 1969, F2.

172. "Quake Alarms, but There Is No Disaster," *Moberly Monitor-Index and Evening Democrat*, April 29, 1969.

174. Susan E. Hough et al., "Sediment-Induced Amplification and the Collapse of the Nimitz Freeway," *Nature* 344 (1990): 853–55.

176. "Experts: Quake Shouldn't Affect New Madrid Fault," *Constitution-Tribune*, Chillicothe, October 18, 1989.

176. "Even Odds Quake Could Hit Illinois within 10 Years," *Daily Herald*, Springfield, Illinois, October 19, 1989.

176–179. James L. Penick, *The New Madrid Earthquakes of 1811–1812* (Columbia: University of Missouri Press).

178. Jared Brooks's accounts of the New Madrid sequence were published as an appendix to: H. McMurtrie, MD, *Sketches of Louisville and Its Environs, Including, among a Great Variety of Miscellaneous Matter, a Florula Louisvillensis; or, a Catalogue of Nearly 400 Genera and 600 Species of Plants, That Grow in the Vicinity of the Town, Exhibiting Their Generic, Specific, and Vulgar English Names* (Louisville: S. Penn., Junior, 1839).

179. Eliza Bryan, letter published in "Lorenzo Dow's Journal," Joshua Martin (publisher) (1816): 344–46.

180–181. Daniel Drake, *Natural and Statistical View, or Picture of Cincinnati and the Miami County, Illustrated by Maps* (Cincinnati: Looker and Wallace, 1815).

181–182. Otto W. Nuttli, "The Mississippi Valley Earthquakes of 1811 and 1812: Intensities, Ground Motion, and Magnitudes," *Bulletin of the Seismological Society of America* 63 (1973): 227–48.

183. Thomas Hanks and Hiroo Kanamori, "A Moment-Magnitude Scale," *Journal of Geophysical Research* 84(1979): 2348–50.

184. Arch C. Johnston and Eugene S. Schweig, "The Enigma of the New Madrid Earthquakes of 1811–1812," *Annual Review of Earth Planetary Sciences* 24 (1996): 339–84.

184. Arch C. Johnston, "Seismic Moment Assessment of Earthquakes in Stable Continental Regions—II. Historical Seismicity," *Geophysical Journal International* 125 (1996): 639–78.

184. Arch C. Johnston, "Seismic Moment Assessment of Earthquakes in Stable Continental Regions—III. New Madrid 1811–1812, Charleston 1886, and Lisbon 1755," *Geophysical Journal International* 126 (1996): 314–44.

184. "Devastating Earthquake 175 Years Ago in Mississippi Valley May Occur Again," *The News*, Frederick, Maryland, December 15, 1986, D8.

185. "Apathy Underestimates Danger of Earthquakes," *Chicago Daily Herald*, November 6, 1987, 9–10.

185. "Climatologist Predicts Missouri Quake," *Atchison Globe*, November 28, 1989, 3.

185. "Scientist's Quake Prediction Creates Stir," *Daily Herald*, October 22, 1990, section 6, page 4.

185–186. Nels Winkless III and Iben Browning, *Climate and the Affairs of Men* (New York: Harper's Magazine Press, 1975).

186. "Scientist's Quake Prediction Creates Stir," *Daily Herald*, October 22, 1990, section 6, page 4.

186. "Quake Prediction Taken Seriously," *St. Louis Post-Dispatch,* July 21, 1990.

186. "If we have to schedule National Guard drills . . .": "Quake Prediction Shakes Up People on New Madrid Fault," *Chillicothe Constitution-Tribute*, July 16, 1990, 14.

186. "Prediction of Major Quake Dec. 2 Has Folks Rattled in the Heartland," *Stars and Stripes*, July 17, 1990, 4.

187. David Stewart, *The Chemistry of Essential Oils Made Simple: God's Love Manifest in Molecules* (Marble Hill, MO: Care Publications, 2004).

187. William Robbins, "Midwest Quake Is Predicted, Talk Is Real," *New York Times*, August 20, 1990.

187. Transcript of proceedings, NEPEC hearing on Browning prediction.

187. Mike Penprase, "Drills Prepare Students in Ozarks for Predicted Quake," *News-Leader*, 1, November 18, 1990.

187. "Haps Tavern . . . In New Madrid, Crowds for the Quake That Wasn't," *New York Times*, December 4, 1990.

187. "Student Absences Not Unusually High," *St. Louis Post-Dispatch*, December 5, 1990.

188. At http://joelarkins.blogspot.com.

188. R. A. Kerr, "The Lessons of Dr. Browning," *Science* 253 (1991): 622–23.

189. S. E. Hough, J. G. Armbruster, L. Seeber, and J. F. Hough, "On the Modified Mercalli Intensities and Magnitudes of the 1811–1812 New Madrid, Central United States, Earthquakes," *Journal of Geophysical Research* 105 (2000): 23839–64.

189. William Bakun and Margaret G. Hopper, "Magnitudes and Locations of the 1811–1812 New Madrid, Missouri, and the 1886 Charleston, South Carolina Earthquakes," *Bulletin of the Seismological Society of America* 94 (2004): 64–75.

190. Donald E. Franklin, "Wind Blast Hurts Two in Family Camping Out to Escape Quake," *St. Louis Post-Dispatch*, 7A, December 4, 1990.

Chapter 15. Measles

Page

191. Jay Leno, *Tonight Show* monologue, May 18, 2008.

191. "Predicting the Quake," *Time* magazine, August 27, 1973.

192. At http://pasadena.wr.usgs.gov/step/.

193. Yosihiko Ogata, "Detection of Precursory Relative Quiescence before Great Earthquakes through a Statistical Model," *Journal of Geophysical Research* 97 (1992): 19845–71.

193–194. Kristi F. Tiampo, John B. Rundle, S. McGinnis, Susanna Gross, and W. Klein, "Pattern Dynamics and Forecast Methods in Seismically Active Regions," *Pure and Applied Geophysics* 159 (2002): 2429–67.

193–194. ". . . quite remarkable": at www.NBCSanDiego.com, October 6, 2004.

194. Alan L. Kafka and John E. Ebel, "Exaggerated Claims about Success Rate of Earthquake Predictions: 'Amazing Success' or 'Remarkably Unremarkable,'" *Eos, Transactions, American Geophysical Union* 86 (52), Fall Meeting Supplement, abstract S43D-03, 2005.

Chapter 16. We All Have Our Faults

Page

196. Guy R. McClellan, The Golden State: A History of the Region West of the Rocky Mountains, Embracing California . . . (Chicago: Union Publishing Company, 1872).

197. Kenji Satake, Kunihiko Shimazaki, Yoshinobu Tsuji, and Kazue Ueda, "Time and Size of a Giant Earthquake in Cascadia Inferred from Japanese Tsunami Records of January 1700," Nature 379 (1996): 246–49; Brian F. Atwater, The Orphan Tsunami of 1700: Japanese Clues to a Parent Earthquake in North America (Seattle: University of Washington Press, 2004).

199. Martitia P. Tuttle et al., "The Earthquake Potential of the New Madrid Seismic Zone," Bulletin of the Seismological Society of America 92 (2002): 2080–89.

199. Andrew Newman, Seth Stein, John Weber, Joseph Engeln, Ailin Mao, and Timothy Dixon, "Slow Deformation and Lower Seismic Hazard at the New Madrid Seismic Zone," Science 23 (1999): 619-21.

199. Balz Grollimund and Mark D. Zoback, "Did Deglaciation Trigger Intraplate Seismicity in the New Madrid Seismic Zone?" Geology 29 (2002): 175–78.

200. Stephen F. Obermeier, Gregory S. Gohn, Robert E. Weems, Robert L. Gelinas, and Meyer Rubin, "Geologic Evidence for Recurrent Moderate to Large Earthquakes near Charleston, South Carolina," Science 227 (1985): 408–11.

200–201. John E. Ebel, "A New Look at the 1755 Cape Ann, Massachusetts Earthquake," Eos, Transactions, American Geophysical Union, abstract S52A-08, 2001.

201. Clarence Dutton, "The Charleston Earthquake of August 31, 1886," U.S. Geological Survey Annual Report 9, 203–528, 1889.

Chapter 17. The Bad One

Page

206. Lindh quote: "Are Recent California Quakes a 'Final Warning,'" New Mexican, July 13, 1992, A6.

206. Kerry E. Sieh and Patrick L. Williams, "Behavior of the Southernmost San Andreas Fault during the Past 300 Years," Journal of Geophysical Research 95 (1995): 6620–45.

207. Edward H. Field et al., "The Uniform California Earthquake Rupture Forecast, Version 2 (UCERF 2)," U.S. Geological Survey Open-File Report 2007-1437, at http://pubs.usgs.gov/of/2007/1437, 2007.

209. David Vere-Jones, "A Branching Model Crack Propagation," Pure Applied Geophysics 114 (1976): 711–25.

209. N. Lapusta and J. R. Rice, "Earthquake Sequences on Rate and State Faults with Strong Dynamic Weakening," EOS Transactions American Geophysical Union 85 (47), Fall meeting supplement, abstract T22A-05, 2004.

211. T. Rockvell, G. Seitz, T. Dawson, and Y. Young, "The Long Record of San Jacinto Fault Paleoearthquakes at Hog Lake: Implications for Regional Patterns of Strain Release in the Southern San Andreas Fault System," abstract, Seismological Research Letters, 77, 270, 2000.

212. Lloyd S. Cluff, Robert A. Page, D. Burton Slemmons, and C. B. Crouse, "Seismic Hazard Exposure for the Trans-Alaska Pipeline," Sixth U.S. Conference and Workshop on Lifeline Earthquake Engineering, ASCE Technical Council on Lifeline Earthquake Engineering, Long Beach, California, August, 2003.

215. Los Angeles El Clamor Público, January 17, 1857, 2, in Duncan Carr Agnew, "Reports of the Great California Earthquake of 1857," reprinted and edited with explanatory notes, at http://repositories.cdlib.org/sio/tech report/50/, 2006.

216. Heaton quote: "How Risky Are Older Concrete Buildings," Los Angeles Times, October 11, 2005.

216. "Retrofitting Tab at County Hospitals over $156 Million," Los Angeles Business Journal, February 4, 2002.

217, 218. Swaminathan Krishnan, Ji Chen, Dimitri Komatitsch, and Jeroen Tromp, "Case Studies of Damage to Tall Steel Moment-Frame Buildings in Southern California during Large San Andreas Earthquakes," Bulletin of the Seismological Society of America, 96, (2006): 1523–37.

221. Lindh quote: "Are Recent California Quakes a 'Final Warning.'"

Chapter 18. Whither Earthquake Prediction?

Page

222. Allen, Transcript of Proceedings, In the Matter Of.

223. Yuehua Zeng and Z.-K. Shen, "Earthquake Predictability Test of the Load Response Ration Method," abstract, Seismological Research Letters, 2008.

223. Niu Fenglin, Paul G. Silver, Thomas M. Daley, Xin Cheng, and Ernest L. Majer, "Preseismic Velocity Changes Observed from Active Source Monitoring at the Parkfield SAFOD Drill Site," Nature 454 (2008), doi:10.1038/ nature07111.

224. Thomas S. Kuhn, *The Structure of Scientific Revolutions* (Chicago: University of Chicago Press, 1996).

224. Olson, Podesta, and Nigg, *The Politics of Earthquake Prediction*.

225–226. Thomas Jordan, personal communication, 2008.

227. Hiroo Kanamori, personal communication, 2008.

228. Clarence Allen, personal communication, 2008.

229. Allen, Transcript of Proceedings, *In the Matter Of.*

General Index

Note: Page numbers in italic type indicate illustrations.

Index of Earthquakes by Year